Elements of
Exterior and Terminal Ballistics

George Klimi

To order additional copies of this book, contact:
Xlibris
844-714-8691
www.Xlibris.com
Orders@Xlibris.com
823676

Elements of
Exterior and Terminal Ballistics

For Ivi, Erio, Dorina

CONTENTS

Chapter 3   127
Standard Atmosphere in Exterior Ballistics

Chapter 4   146
Elementary Exterior Ballistics

Chapter 5   202
Differential Equations of Exterior Ballistics

Second Part: Elements of Terminal Ballistics

<u>Chapter 6</u>   241
Exterior Ballistics of Fragments

AUTHOR

George Klimi (Gjergj Klimi), Ph.D., is a retired associate professor of Mathematics who taught Math at NYC College of Technology and at Pace University (New York).

George is the author of five books on exterior ballistics: "Exterior Ballistics with Applications" (2008), "Exterior Ballistics of Small Arms" (2009), Exterior Ballistics: A New Approach (2010), Exterior Ballistics: The Remarkable Methods, (2014) Elements of Exterior Ballistics: Long Range Shooting (2016).
"Exterior Ballistics with Applications" was updated in December 2011.

George Klimi, from 1970 until 1993, has been professor of Physics & Mathematics and chair of Physics Department at the Military Academy of Tirana, Albania.
In 90's George used to work at the Committee of Science and Technology, and at the Ministry of Higher Education and Research in Albania as Director of Tempus Office, responsible for the implementation of the European Community Programs to restructure the higher education.

He was awarded the Gold Medal of Eagle, by the President of Republic of Albania, for his contribution in the democratic movement against dictatorship, and for restructuring the higher education in Albania.

PREFACE

"Elements of Exterior and Terminal Ballistics" includes the exterior ballistics of projectiles and the terminal ballistics related to the action of munition fragments at the impact point (soft or hard targets).

As a natural extension of exterior ballistics, the terminal ballistics uses the exterior ballistics to predict the trajectory of metallic fragments set in motion during the detonation of explosive charge of munition.

The book, we are presenting, is a collection of works I have done in exterior ballistics and terminal ballistics.
The book contains two parts:

- Elements of Exterior Ballistics
- Elements of Terminal Ballistics

The readers with a good background in Pre-calculus and Calculus have no difficulty to comprehend the topics presented in the book. In addition, the large number of examples, in each topic, demonstrates the solving techniques of problems and helps the reader to understand the concepts and principles, as well as the challenging theoretical and practical applications.

Each topic is methodically and rigorously presented to avoid ambiguity that is present in the contemporary exterior and terminal ballistics, especially in non-mathematical, or non-technical presentation of exterior ballistics.

The first part of the book, Exterior Ballistics, is the second edition of the "Elements of Exterior Ballistics: Long Range Shooting", G. Klimi, Xlibris 2016.

The second part, Elements of Terminal Ballistics, is an improved and upgraded edition of chapter 10, published at "Exterior Ballistics with Applications", G. Klimi, Xlibris 2011.

The second edition of "Elements of Exterior Ballistics: Long Range Shooting" have corrections of the errors that came across reading the book.
Secondly, we added new problems and topics to enrich the content of the book.

In most of examples we use the 0.338 Lapua GB528 Scenar 19.44g bullet to illustrate and develop different models and some new original Exterior Ballistics methods.
This particular bullet has a lot of information literature that is necessary to prove and support our findings.

George Klimi, PhD
Clearwater, Florida
January 2021
iven24@aol.com,

Terminology

EBA, Exterior Ballistic with Application
EBNA, Exterior Ballistics A New Approach
EBRM, Exterior Ballistics: Remarkable Methods
EBLR, Exterior Ballistics: Long Range Shooting
EETB, Elements of Exterior and terminal Ballistics
ICAO, International Civil Aviation Organization Atmosphere
ASM, Army Standard Metro Atmosphere
TSA, Traditional Standard Atmosphere
BC, Ballistic Coefficient
CD, Drag Coefficient
LOS, Line of Sight
NIH, Natural Irregular Fragments
HER, Equivalent Horizontal Range

DISCLAIMER

Neither the author nor Xlibris accepts any responsibility, or liability for any problem (errors, damages, injuries, etc.) that might occur in practice of shooting using the information given in this book, and by employing the PC programs to solve practical ballistics problems.

The user of the ballistics methods and the associated PC programs that are presented in the book, should be cautious, and use the common sense to judge the predicted outcomes and their practical use.

There might be as well technical errors. It is impossible to find qualified proofreaders in Exterior and Terminal Ballistics.

First Part

Elements of Exterior Ballistics
Second Edition

1

A Brief Approach to Aiming with Small Arms

Introduction

This chapter contains the methods related with zeroing in a rifle at a given range.

There is shown a mathematical relationship between the coordinate system, related with the horizontal range, as well as the rectangular system of coordinates related with the line of sight (LOS).

There are presented the relations between projectile drop, departure angle and bullet path. Those relations are mathematically described by simple equations.

There are shown, as well some methods we use to change the zeroing in.

1.1 Projectile Ballistic Trajectory

To study the flight of projectiles, the traditional exterior ballistics (EB) employs the simplified mathematical model of point-mass. The point-mass model evolves in a system of differential equations that describes the projectile trajectory.

Point-Mass Model

- The projectile is considered a dimensionless body with mass concentrated at the center of mass of the projectile.
- The only forces that act on the point-mass projectile during the flight, include the drag force and gravity.
- The point-mass projectile considers as well the influence of atmospheric conditions (temperature, pressure, density of air, air humidity).
- The initial conditions (departure velocity, launching angle), at the departure point of projectile, predict the position, velocity, angle of the projectile at any time along the flight path.
- To compensate for the missing dimensions of the point-mass of the projectile EB introduces the well-known ballistics coefficient (BC) of the projectile.

Using the point mass model, we obtain a system of differential equations (section 5.1) that describes the ballistics trajectory of projectile.

The ballistics trajectory is the flight course of an unguided point-mass projectile, taken under the action of drag force and gravity, after the projectile leaves the muzzle of the firearm with a given initial velocity till it hits the target, or the ground.

The influence of wind, projectile rotation, rotation of the Earth, etc. are either ignored (when their results are insignificant), or introduced in the point-mass model to adjust the projectile trajectory.

The exterior ballistics predicts the ballistics trajectory with respect to a three-dimensional **system of Cartesian coordinates** xoyz that has the origin at the muzzle of the firearm, while x-axis is along the horizontal line.
The direction of x-axis is along the projection of the initial velocity $\vec{v}_0$ on the horizontal line.

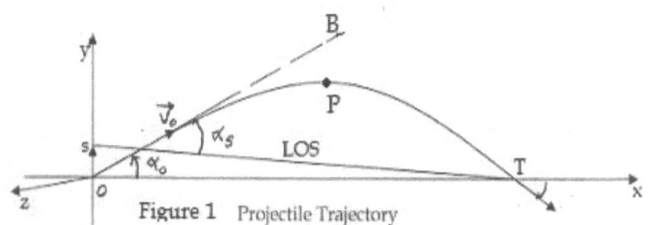

Figure 1   Projectile Trajectory

In general, the muzzle of the firearm is at the point with coordinates $(0, y_0)$, i.e. at the altitude y_0 over the sea level, or over the firing site.

The ballistic trajectory of a point-mass projectile is plane, i.e. the flying projectile does not deviate from the launching plane xoy (there is no crosswind, or any other lateral deviation due to the rotation of the projectile, etc.).

The ballistics trajectory is a set of points with coordinates (x, y, z = 0). The coordinates of the projectile are functions of time of flight (t), i.e. $x = x(t)$, $y = y(t)$, $z = z(t) = 0$.
At any point (x, y), that corresponds to the time of flight t, the projectile has a defined velocity $\vec{v}$ tangent to the trajectory, i.e. a defined speed v and an angle α that the velocity forms with x-axis.
The projectile trajectory has the following characteristic elements:
The departure angle α_0, the initial velocity (muzzle speed) v_0, the ballistics coefficient c, the horizontal range $x_T = OT$ (or the

coordinates of the point of impact (x_T, y_T, $z_T = 0$), the time of flight to the target t_T, the terminal speed v_T, the angle of impact on the target α_T and the maximum trajectory altitude y_m (or the coordinates of the trajectory vertex (x_m, y_m, $z_m = 0$).

The target (T) can be at any point over, under, or at the altitude of the firearm.

Initial Velocity and Exit Velocity of Projectile
When a round is fired, the gases of the propellant charge exit the muzzle of the firearm at a velocity greater than the bullet exit velocity (for some bullets around 50% greater).

The hot and high-pressure gases, expanding and interacting with the bullet, increase the projectile velocity beyond the exit velocity at the muzzle, until the bullet travels some centimeters to some few meters (that depend on the type of projectile (bullet), propellant charge and firearm).

At the end of this transition period, the bullet has reached the maximum velocity (this velocity is not the initial velocity).

The **initial velocity** of a bullet (projectile) is a **fictive velocity** that let the projectile follow the real trajectory. The projectile is launched with the **fictive velocity**, assuming that at the muzzle the powder gases cease to increase the velocity of bullet beyond the muzzle.

For the initial velocity there are in use terms: departure velocity, muzzle velocity, departure speed.

The muzzle velocity of a standard projectile, fired from a brand new barrel, depends on the propellant temperature.

For the ICAO atmosphere, and for the ASM atmosphere, the standard temperature of propellant charge is 21.11 degree Celsius (70 degree Fahrenheit).

The standard temperature of the propellant charge for the TSA atmosphere is 15 degree Celsius (59 degree Fahrenheit).

The propellant temperature, which is usually equal to the temperature of air at the shooting site, changes the initial standard velocity of the projectile when the temperature of air at the firing site is different from the temperature of the standard atmosphere. For that reason, some standard range tables, which are valid for firing in a standard atmosphere, contain range corrections related to the change in propellant temperature.

Initial Velocity and Temperature of Cartridge
The actual initial velocity of a projectile V_0, for a given temperature T of propellant charge, can be estimated using the equation

$$V_0 = v_0 \cdot [1 + 0.001 \cdot (T - T_0)], \qquad (1.1.1)$$

where v_0 is the initial standard velocity of the projectile that corresponds to the standard temperature T_0 of propellant (21.11 degree Celsius, for ICAO and ASM atmospheres; 15 degree for TSA atmosphere).

Initial Velocity and Bullet Mass
For the same cartridge and propellant charge, any change in bullet mass, however small, changes the initial velocity of the bullet from v_0 to

$$v_{0m} = v_0 (1 - 0.4 \frac{dm}{m_0}), \qquad (1.1.2)$$

where m_0 is the standard mass of bullet, $dm = m - m_0$ is the change in bullet mass.

Thus, if the standard initial velocity of a bullet is 850 m/s, and if the relative increase of bullet mass is $dm/m_0 = 0.01 = 1\%$, then there is a change in the initial velocity from 850m/s to

$$v_{0m} = v_0(1 - 0.4\frac{dm}{m_0}) = 850 \cdot (1 - 0.4 \cdot 0.01) = 846m/s .$$

Note that both formulas (1.1.1) and (1.1.2) are considered in the PC programs associated with the book EBRL.

Initial Velocity and Barrel Length
The muzzle velocity of a given bullet changes with the change of barrel length.
The change in muzzle velocity Δv_0 with a change ΔL_0 in barrel length can be estimated approximately by the equation (See Exterior ballistics: The Remarkable Method, page 455)

$$\Delta v_0 = 0.28916 \cdot v_0 \cdot \Delta L_0 . \tag{1.1.3}$$

Hence, for the velocity that corresponds to the increased or decreased length of barrel, we have

$$v_{0B} = v_0[1 + 0.28916 \cdot (L - L_0)] . \tag{1.1.4}$$

Thus, if the departure velocity of a bullet is 830m/s then for an increase of $\Delta L_0 = L - L_0 = 0.05m$ in barrel length the muzzle velocity will be

$$v_{0B} = v_0[1 + 0.28916 \cdot (L - L_0)] = 830 \cdot (1 + 0.28916 \cdot (0.05) = 842m/s.$$

Example 1.1

The standard initial velocity of a projectile, fired in ICAO atmosphere, is 860m/s (if we consider that the projectiles are stored in a temperature 21.11 degree Celsius).

In shooting situations, it is usually not possible to keep the temperature of propellant charge equal to 21.11 degree Celsius.

Assume that the actual temperature of air at the firing site is 10 degree Celsius.

What will be the departure velocity of the projectile if the temperature of projectile cartridge (propellant charge) is the same as the actual temperature of air?

Solution

Using (1.1.1), we find that the departure velocity of the projectile is

$$V_0 = v_0 \cdot (1 + 0.001 \cdot (T - T_0)) = 860 \cdot (1 + 0.001 \cdot (10 - 21.11)) = 850.54 \ m/s$$

The change in initial velocity is

$$\Delta(v_0) = V_0 - v_0 = 850.54 - 860 = -9.46 \ m/s.$$

1.2 Trajectory, Bullet Drop, Departure Angle

Geometric Elements of Ballistic Trajectory

The geometric elements of the ballistic trajectory of a projectile, with respect to a rectangular Cartesian system of coordinates are (see fig. 1 and fig 2):

- The horizontal line (ox) is the line that originates at the muzzle of the firearm and has the direction of the horizontal component of initial velocity.
- Line of Departure is the tangent line to the ballistic trajectory at the muzzle of the firearm and has the direction

of the initial velocity ($\vec{v}_o$) of the projectile. The initial speed of the projectile is v_o.

It is evident that the departure line is the direction of the axis of the muzzle   of firearm just as the projectile leaves the muzzle.

- Angle of departure (α_0) is the angle measured from x-axis to the line of departure.
- Line of sight (LOS) is the straight line that connects the muzzle of the firearm and the target (center of the target, or a point on the target).

  For small arms or sight firearms, the LOS is the line that connects the eye of the marksman, the scope, and the target. LOS is neither parallel to the axis of the bore, nor to the horizontal line (x-axis).

  Note that the direction of LOS changes when we change the horizontal range of the target.
- Aiming angle is the angle measured from LOS to the departure line LOD=OB.
- In inclined shooting, the elevation/depression angle is the angle measured from the x-axis to the line that connects the origin of coordinates (muzzle of firearm) and the target located on the inclined plane.
- Super elevation angle is the angle that zeroes the rifle on the inclined range
- The **horizontal range** $x_T = OT$ is the distance from the muzzle of the firearm to the target T, when both the target and the muzzle of firearm are at the same altitude over the sea level ($y_0 = y_T = 0$).
- Projectile drop, at a given point P on the trajectory, is the vertical distance QP from the line of departure to the point P (fig. 2).

Bullet Drop

The drop of a projectile $\bar{y}_D = QP$, at a given time and location P, during the flight, is the perpendicular distance of bullet measured from the line of departure OB to the point P on the trajectory (fig. 2).

The projectile drop, in the coordinate system xoy, is a negative number, i.e. $\bar{y}_D < 0$

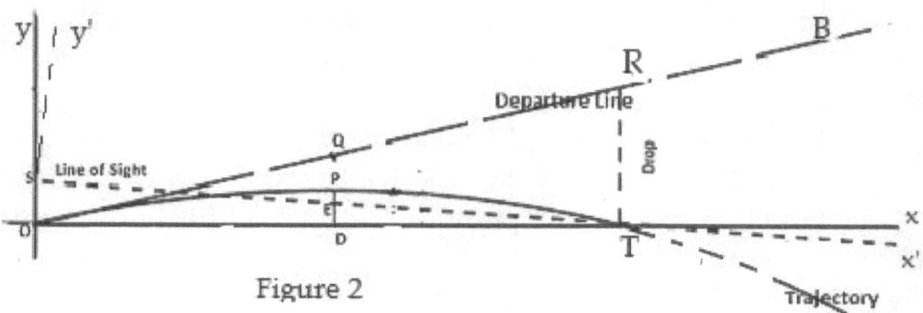

Figure 2

Departure Angle

In some standard range tables, it is given the bullet drop $\bar{y}_T$ at zero-range $x_T = OT$. Using the bullet drop, $\bar{y}_T$ we can find the departure angle α_{0T} that zeroes in the firearm at horizontal range $x_T = OT$.

The projectile drop that corresponds to the horizontal range $x_T = OT$ is $\bar{y}_T = RT$ The departure angle $\alpha_{0T} = \angle TOB$ is the departure angle needed to hit the target located at the horizontal range $x_T = OT$.(fig. 1, fig. 2).

In other words, α_{0T} is the departure angle that zeroes the firearm at $x_T = OT$.

From fig. 2, for the departure angle α_{0T} , we can write:

$$\tan \alpha_{0T} = \frac{|\bar{y}_T|}{x_T} = -\frac{\bar{y}_T}{x_T} \qquad (1.2.1)$$

Hence,

$$\alpha_{OT} = \arctan(\frac{|\bar{y}_T|}{x_T})$$ (1.2.2)

Note that the bullet drop is a negative number, $\bar{y}_T < 0$.
Since in the long range shooting the departure angles are usually very small, we can write equation (1.2.1) as

$$\alpha_{OT} \approx \frac{|\bar{y}_T|}{x_T} \cdot \frac{180}{\pi},$$ (1.2.3)

where
α_{OT} is measured in degree.
For the departure angle in MOA (Minutes of Angle), using equation (1.2.3) we can write:

$$\alpha_{OT} \approx \frac{|\bar{y}_T|}{x_T} \cdot \frac{10,800}{\pi}.$$ (1.2.4)

Bullet Drop and Departure Angle
Using (1.2.1), for the projectile drop (in absolute value), at the horizontal range $x_T = OT$, we have:

$$|\bar{y}_T| = x_T \cdot \tan\alpha_{OT}$$ (1.2.5)

or, since the drop is negative,

$$\bar{y}_T = -x_T \cdot \tan\alpha_{OT}.$$ (1.2.6)

To find the drop $\bar{y}_D = QP$ of the projectile at any point P on the trajectory, when the projectile is lunched at the angle α_{OT} that zeroes the firearm at $x_T = OT$, we need to know the height

$y_D = DP$ of the trajectory at the point P with abscissa x_D. The projectile drop, $\bar{y}_D = QP$, at the point P with abscissa x_D (fig. 2) is

$$\bar{y}_D = y_D - y_Q = y_D - x_D \tan\alpha_{OT}, \qquad (1.2.7)$$

Where y_D is the y-coordinate of the projectile located at a point P on the trajectory.

The quantity y_D is at the same time the height of bullet over the horizontal line ox (when bullet is launched with angle α_{OT} that zeroes the firearm at $x_T = OT$).

Dividing both sides of (1.2.7) by x_D, we have:

$$\frac{\bar{y}_D}{x_D} = \frac{y_D}{x_D} - \tan\alpha_{OT}. \qquad (1.2.8)$$

Note that the range, $x_D = OD$, can be smaller or greater than the zero-range $x_T = OT$.

The left side of (1.2.8) is the tangent of the departure angle α_{0D} that zeroes the rifle at the horizontal range $x_D = OD$

Thus, for the departure angle α_{0D} that zeroes in the rifle at $x_D = OD$, we have:

$$-\tan\alpha_{0D} = \frac{y_D}{x_D} - \tan\alpha_{OT}. \qquad (1.2.9)$$

Hence,
$$\tan\alpha_{0D} = \tan\alpha_{0T} - \frac{y_D}{x_D}, \qquad (1.2.10)$$

or approximately

$$\alpha_{0D} \approx \alpha_{0T} - \frac{y_D}{x_D} \cdot \frac{180}{\pi} \qquad (1.2.11)$$

Hence, for the change in departure angle, when there is a change in zero range, we have:

$$\alpha_{0D} - \alpha_{0T} \approx -\frac{y_D}{x_D} \cdot \frac{180}{\pi}. \qquad (1.2.12)$$

Denoting

$$\Delta\alpha_{OT} = (\alpha_{0D} - \alpha_{OT}), \qquad (1.2.13)$$

the change in respective ranges, we can write (1.2.12), in the following form:

$$\Delta\alpha_{OT} \approx -\frac{y_D}{x_D} \cdot \frac{180}{\pi}. \qquad (1.2.14)$$

NOTES
- **Radian and Degree Measures**
  In international system of units (SI), the angles are measured in radian.
  In the book we measure the angles mostly in degree.
  To convert an angle measured in radian into degree, we multiply the angle by $(180/\pi)$ and, vice versa, by $(\pi/180)$

- **MOA**
  Another unit used in long range shooting is a minute of angle (MOA).
  To convert an angle measured in degree into MOA we multiply by 60 the angle in degree.
  To convert an angle measured in radian into an equal angle measured in degree we use the multiplication factor ($180/\pi$).
  To convert an angle measured in radian into an angle measured in MOA we use the multiplication factor ($10,800/\pi$).

- **Approximate Trigonometric Equations**

Throughout the book, since the departure angles in long range shooting   with small arms are relatively small, close to zero degree, we use the   approximate equations:

(a) For angle α measured in radian

$$\sin \alpha \approx \alpha, \qquad \tan \alpha \approx \alpha, \qquad\qquad (1.2.15)$$

(b) For angle α measured in degree

$$\sin \alpha \approx \alpha \cdot \frac{\pi}{180}, \qquad \tan \alpha \approx \alpha \cdot \frac{\pi}{180}. \qquad (1.2.16)$$

Thus, for example, the solution of the equation

$$\sin \alpha = 0.0262$$

is

$$\alpha = \arcsin(0.0262) = 1.50132°.$$

Employing (1.2.15), for the approximate solution we can write:

$$\alpha \cdot \frac{\pi}{180} \approx \sin \alpha = 0.0262.$$

Hence,

$$\alpha \approx 0.0262 \cdot \frac{180}{\pi} = 1.50115°.$$

See as well, Example 2.4, section 1.2:

$$\alpha_{0D} \approx \alpha_{0T} - \frac{y_D}{x_D} \cdot \frac{180}{\pi} = 0.5247 - \frac{68.70}{800 \cdot (36)} \cdot \frac{180}{\pi} = 0.3380°.$$

- **Converting Inch/yard to meter/meter**

$$\frac{1\ Inch}{1\ yard} = \frac{1 \cdot 0.0254\ m}{1 \cdot 0.9144\ m} = \frac{1}{36}$$

Hence,

$$1\ yard = 36\ inch$$

Thus, for example

$$\frac{2.7\ Inch}{100\ yard} = \frac{2.7}{100} \cdot \frac{1}{36} = 0.075$$

See as well, Example 2.4, section 1.2:

$$\alpha_{0D} \approx \alpha_{0T} - \frac{y_D}{x_D} \cdot \frac{180}{\pi} = 0.5247 - \frac{68.70}{800 \cdot (36)} \cdot \frac{180}{\pi} = 0.3380°.$$

where
y_D=68.70 inches, x_D=800 yards.

Example 2.1 Departure Angle
The drop of 0.338 Lapua GB528 Scenar 19.44g bullet at the horizontal range $x_T = 1{,}500m$ is $\bar{y}_T = -30.035m$.
Find the departure angle needed to zero the firearm at the given horizontal range.

Solution
Employing (1.2.1), we find that

$$\alpha_{OT} = -\arctan(\frac{\bar{y}_T}{x_T}) = -\arctan(\frac{-30.035}{1500}) = 1.1471°.$$

Employing (1.2.3), we get quite the same departure angle

$$\alpha_{OT} = -\frac{\bar{y}_T}{x_T} \cdot \frac{180}{\pi} = -(\frac{-30.035}{1500}) \cdot \frac{180}{\pi} = 1.1473°.$$

Example 2.2 Projectile Drop
The rifle, firing a 0.338 Lapua GB528 Scenar 19.44g bullet, is zeroed in at the horizontal range, $x_T = 1,200$ meters. The departure velocity of bullet is 830m/s (2723.10 fps).
According to table 3.1 section 5.3, the departure angle that zeroes the rifle at $x_T = 1,200$ meters is $\alpha_{0T} = 0.7838°$.
Find the projectile drop at the horizontal range.

Solution
Substituting in (1.2.3) we find that the bullet drop is

$$\bar{y}_T = -x_T \cdot \tan\alpha_{0T} = -1200 \cdot \tan(0.7838) = -16.42m.$$

Example 2.3 Projectile Drop
The rifle firing a 0.338 Lapua GB528 Scenar 19.44g bullet is sighted in at the horizontal range $x_T = 1,200$ meters. The departure angle that zeroes the rifle at $x_T = 1,200$ meters is $\alpha_{0T} = 0.7838°$ (see table 3.1, section 5.3). The departure velocity of bullet is 830 m/s.
Find the projectile drop at the range $x_D = 900$ meters, if the height of the projectile over the horizontal line is $y_D = 4.24m$.

Solution
Using (1.2.7), we find that the projectile drop is

$$\bar{y}_D = y_D - x_D \cdot \tan\alpha_{0T} = 4.24 - 900 \cdot \tan(0.7838) = -8.072m.$$

Note
The drop, estimated above, is the same as the drop calculated when the rifle is zeroed at 900 meters.
Indeed, the departure angle that zeroes the rifle at $x_D = 900$ meters is $\alpha_D = 0.5126°$ (see table 3.1, section 5.3). Employing (1.2.6) we find the drop:

$$\bar{y}_D = -x_D \cdot \tan\alpha_D = -900 \cdot \tan(0.5126) = -8.052m$$

Example 2.4 Predicting Departure Angle

A rifle fires a 0.338 Lapua GB528 Scenar 19.44g bullet, with departure velocity 2723 fps in ICAO atmosphere.

The departure angle that zeroes the rifle at $x_T = 1,000$ yards, is $\alpha_{0T} = 0.5247°$.

At $x_D = 800$ yards, the height of the bullet trajectory above horizontal line is $y_D = 68.70$ inches.

Find the departure angle α_{0D} that zeroes the firearm at $x_D = 800$ yards.

Solution

Substituting in (1.2.11), we find that the departure angle that zeroes the rifle at $x_D = 800$ yards is

$$\alpha_{0D} \approx \alpha_{0T} - \frac{y_D}{x_D} \cdot \frac{180}{\pi} = 0.5247 - \frac{68.70}{800 \cdot (36)} \cdot \frac{180}{\pi} = 0.3380° \, .$$

Example 2.5

The rifle firing a 0.338 Lapua GB528 Scenar 19.44g bullet with velocity 830 m/s, is zeroed at the horizontal range $x_{0T} = 600$ meters.

At the horizontal distance $x_D = 300$ meters the height of the trajectory above the horizontal line is $y_D = 0.872m$.

What is the change in departure angle if we change the zero range from $x_{0T} = 600$ to $x_D = 300$ meters?

Find the change in departure angle.

Solution

Using (1.2.12), we find that the change in departure angle is

$$(\alpha_{0D} - \alpha_{0T}) \approx -\frac{y_D}{x_D} \cdot \frac{180}{\pi} = -\frac{0.872}{500} \cdot \frac{180}{\pi} = -0.0999°.$$

1.3 Trajectory Height above the Horizontal Line

The trajectory height $y_D = DP$ above the horizontal line (ox) is usually given in standard range tables.
It can be obtained solving the system of differential equations (2.5.1), or using other methods.

Let's estimate the projectile height over (or below) the horizontal line ox. We assume that the rifle is zeroed in at range $x_T = OT$ and α_{0T} is the departure angle corresponding to the zero-range $x_T = OT$. If the projectile height $y_D = DP$, at the horizontal range $x_D = OD$ is not known, then using (1.2.7), for the y-coordinate of the projectile height at the given range, x_D we have:

$$y_D = x_D \tan \alpha_{0T} + \bar{y}_D, \tag{1.3.1}$$

where drop $\bar{y}_D < 0$. The equation (1.3.1) can be written:

$$y_D = x_D \cdot \frac{y_T}{x_T} + \bar{y}_D. \tag{1.3.2}$$

Let's consider that we know the angles $\alpha_{0T} = \angle TOA$ and $\alpha_{0D} = \angle DOP$ that zero the firearm respectively at $x_T = OT$ and $x_D = OD$ (fig. 3 below)
The projectile height y_D at the range $x_D = OD$, when the firearm is zeroed at the range $x_T = OT$, is (fig. 2):

$$y_D = x_D \tan\alpha_0 - x_D \tan\alpha_D = x_D(\tan\alpha_{0T} - \tan\alpha_{0D}) . \qquad (1.3.3)$$

Using equation (1.3.3), we can write the equivalent equation:

$$y_D = x_D(\frac{|\bar{y}_T|}{x_T} - \frac{|\bar{y}_D|}{x_D}) . \qquad (1.3.4)$$

Equation (1.3.3) can be written in approximate form as:

$$y_D \approx x_D(\alpha_{0T} - \alpha_{0D}) \cdot \frac{\pi}{180} . \qquad (1.3.5)$$

Note that the point D can be in front of point T, or behind T, respectively when range $x_D = OD$ is smaller, or greater than zero-range $x_T = OT$.

Example 3.1 Trajectory height over the horizontal line
The rifle firing a 0.338 Lapua GB528 Scenar 19.44g bullet is zeroed at the horizontal range $x_{OT} = 1{,}000$ meters. The corresponding departure angle that zeroes the rifle at 1,000 meters is $\alpha_{0T} = 0.5958°$ (see table 3.1, section 5.3).
(a) Find the height of the bullet trajectory at $x_D = 900$ meters.
(b) Find the bullet height at range $x_D = 1{,}100$ meters.

Solution
(a) In table 3.1 section 5.3, we find that the departure angle that zeroes the firearm at $x_D = 900$ meters is $\alpha_{0D} = 0.5126°$.
Employing equation (1.3.3), we find that the bullet (at 900 meters) passes

$$y_D = x_D(\tan\alpha_{0T} - \tan\alpha_{0D}) = 900 \cdot (\tan 0.5958 - \tan 0.5126) = 1.31m$$

over the horizontal line (fig. 3).

(b) In table 3.1, section 3.5, we find that the departure angle that zeroes the firearm at $x_D = 1{,}100$ meters is $\alpha_{OD} = 0.6858°$. Employing equation (1.3.3) we have:

$$y_D = x_D(\tan\alpha_{OT} - \tan\alpha_{OD}) = 1100 \cdot (\tan 0.5958 - \tan 0.6858) = -1.728m$$

The y-coordinate (height) of bullet trajectory at $x_D = 1{,}100$ meters, when the firearm is zeroed at $x_{OT} = 1{,}000$ meters, is negative, i.e. the bullet trajectory is under the horizontal line.

We get the same value using (1.3.5). Indeed,

$$y_D \approx x_D(\alpha_{OT} - \alpha_{OD}) \cdot \frac{\pi}{180} = 1100 \cdot (0.5958 - 0.6858) \cdot \frac{\pi}{180} = -1.728.$$

Example 3.2 Trajectory height
A shooter with a Russian SKS rifle fires a 7.62mm bullet with departure speed 735m/s.
At what point over the horizontal line the bullet will hit a rectangular table-target located at a distance of $x_D = 100$ meters from the SKS in order that the firearm must be zeroed at $x_T = 300$ meters?

From the standard range table of the Russian rifle we find that the drop of the bullet at $x_D = 100$ meters and $x_T = 300$ meters is respectively $\bar{y}_D = -0.15m$ and $\bar{y}_T = -1.13m$.

Solution

Aiming at $x_T = 300$ meters, with the corresponding scope sight the gunman fires some bullets.

In accordance with equation (1.3.4), we expect that, at $x_T = 300$ meters, the vertical deviation of center of distributions of bullets from the bottom of the rectangular board-target to be

$$y_D = x_D(\frac{|\bar{y}_T|}{x_T} - \frac{|\bar{y}_D|}{x_D}) = 100 \cdot (\frac{1.13}{300} - \frac{0.15}{100}) = 0.227m .$$

1.4 Aiming Angle, Angle of Sight-Scope, Departure Angle

Aiming Angle α_S is the angle measured from the LOS = ST to the departure line OB.

Angle of Sight-Scope is the angle $\alpha_h = \angle OTS$ measured from the horizontal line OT to LOS (fig., 1, fig. 2 above).

The **angle of the sight-scope is negative** since it is obtained by a clock-wise rotation.

Relationship Between Aiming Angle, Departure Angle and Angle of Sight-Scope

Assume that the rifle is zeroed in at range $x_T = OT$.

If the scope height of the rifle is $y_S = h_T$, and we aim at $x_T = OT$ then the **angle of sight-scope,** $\alpha_h = \angle OTS$, is given by the equation:

$$\tan \alpha_{hT} = -(h_T / x_T) , \qquad (1.4.1)$$

or, approximately by

$$\alpha_{hT} \approx -\frac{h_T}{x_T} \cdot \frac{180}{\pi} . \qquad (1.4.2)$$

The **departure angle** $\alpha_{OT} = \angle TOR$ zeroes the rifle at range $x_T = OT$. The aiming angle α_{ST} is given by the formula:

$$\alpha_{ST} = \alpha_{OT} - \alpha_{hT},$$ (1.4.3)

which can be written in compact form:

$$\alpha_{ST} \approx (\frac{|\overline{y_T}|}{x_T} + \frac{h_T}{x_T}) \cdot \frac{180}{\pi}$$ (1.4.4)

Example 4.1
A Sierra NATO bullet caliber 0.30, 168 grain HPBT, fired horizontally, at the sea level, with initial speed 2,650 fps at the horizontal range $x_T = 600$ yards.

At $x_T = 600$ yards, the bullet drop of is $\overline{y}_T = 126.50$ inches.

(a) Find the angle of departure needed to zero the gun at a horizontal range of $x_T = 600$ yards from the gun.

(b) Find as well the aiming angle that the line of bullet departure forms with the line of sight (LOS). The sight-scope height is 1.5 inches.

Solution
(a) The departure angle α_{OT} that zeros the firearm at the horizontal range $x_T = 600$ yards is

$$\alpha_{OT} = \frac{|\overline{y}_T|}{x_T} \cdot \frac{180}{\pi} = \frac{126.50}{600 \cdot (36)} \cdot \frac{180}{\pi} = 0.3356° = 20.133 MOA.$$

(b) For the angle of sight (aiming angle) that the line of sight (LOS) forms with the horizontal line at the horizontal range 548.60m, we can write:

$$\alpha_{hT} = -\frac{h_T}{x_T} \cdot \frac{180}{\pi} = -\frac{1.5}{600 \cdot (36)} \cdot \frac{180}{\pi} = -0.00398° = 0.2387 MOA.$$

The angle of sight is

$$\alpha_{ST} = \alpha_{0T} - \alpha_{hT} = 0.3356 - (-0.00398) = 0.3396° = 20.37 MOA.$$

Compact Solution
Substituting in (1.4.4), for the aiming angle we have:

$$\alpha_{ST} \approx \left(\frac{|y_T|}{x_T} + \frac{h_T}{x_T}\right) \cdot \frac{180}{\pi} = \frac{126.50 + 1.5}{600 \cdot 36} \cdot \frac{180}{\pi} = 0.3395°$$

1.5 Change of Zero Range

For small changes, $\Delta\alpha_{0T}$ and $\Delta\alpha_{hT}$, that correspond respectively to departure angle α_{0T} and to angle of sight-scope α_{hT}, the aiming angle α_{ST} (defined in formula (1.4.3)) changes with the quantity given by equation

$$\Delta\alpha_{ST} = \Delta\alpha_{0T} - \Delta\alpha_{hT}, \tag{1.5.1}$$

where $\Delta\alpha_{hT} < 0$,

$$\Delta\alpha_{0T} = \alpha_{OD} - \alpha_{OT}, \tag{1.5.2}$$

and

$$\Delta\alpha_{hT} = \alpha_{hD} - \alpha_{hT}. \tag{1.5.3}$$

Thus, the change in aiming angle is

$$\Delta\alpha_{ST} = (\alpha_{OD} - \alpha_{OT}) - (\alpha_{hD} - \alpha_{hT}). \tag{1.5.4}$$

Compact Equivalent Equation

Let's express the above equations through the drops and the angles of sight scopes.

Assume that the rifle is zeroed at a given range x_T and the corresponding departure angle is

$$\alpha_{0T} = \frac{|\bar{y}_T|}{x_T} \cdot \frac{180}{\pi}, \qquad (1.5.5)$$

The angle of sight-scope is

$$\alpha_{hT} = -\frac{h_T}{x_T} \cdot \frac{180}{\pi}. \qquad (1.5.6)$$

Aiming angle at x_T is

$$\alpha_{ST} \approx (\frac{|\bar{y}_T|}{x_T} + \frac{h_T}{x_T}) \cdot \frac{180}{\pi}. \qquad (1.5.7)$$

Note that h_T is the sight-scope height when the zero range is x_T. Without changing the sight-scope height h_T, we change the LOS by aiming at the point D that is at the range $x_D = OD$.

At the same time, we change the departure angle from α_{0T} to α_{0D} where α_{0D} correspond to range $x_D = OD$.

Though the scope height does not change, the angle of sight-scope changes from the value α_{hT} given in (1.5.6) to the value

$$\alpha_h = -\frac{h_T}{x_D} \cdot \frac{180}{\pi}. \qquad (1.5.8)$$

Thus, the new aiming angle is

$$\alpha_{SDT} \approx (\frac{|\bar{y}_D|}{x_D} + \frac{h_T}{x_D}) \cdot \frac{180}{\pi}. \qquad (1.5.9)$$

The actual change in aiming angle is

$$\Delta\alpha_{SDT} \approx [(\frac{|\bar{y}_D|}{x_D} - \frac{|\bar{y}_T|}{x_T}) + (\frac{h_T}{x_D} - \frac{h_T}{x_T})] \cdot \frac{180}{\pi}. \qquad (1.5.10)$$

Now, we keep unchanged the first term of (1.5.10) and adjust the height of sight-scope at the new zero-range $x_D = OD$. The height of sight-scope at new zero range is denoted h_D.
As result of two successive operations, the total change in aiming angle is

$$\Delta\alpha_{ST} \approx [(\frac{|\bar{y}_D|}{x_D} - \frac{|\bar{y}_T|}{x_T}) + (\frac{h_D}{x_D} - \frac{h_T}{x_T})] \cdot \frac{180}{\pi}. \qquad (1.5.11)$$

where

$$\alpha_{hD} = -\frac{h_D}{x_D} \cdot \frac{180}{\pi}. \qquad (1.5.12)$$

Thus (1.5.11) estimates the correction in aiming angle needed to zero the firearm at the zero-range $x_D = OD$.
The change in aiming angle, given in (1.5.11), in practice is done by changing the sight-scope angle with the quantity

$$\Delta\alpha_h = \Delta\alpha_{ST}. \qquad (1.5.13)$$

Thus, the change in sight-scope angle is

$$\Delta\alpha_{hT} \approx [(\frac{|\bar{y}_D|}{x_D} - \frac{|\bar{y}_T|}{x_T}) + (\frac{h_D}{x_D} - \frac{h_T}{x_T})] \cdot \frac{180}{\pi}. \qquad (1.5.14)$$

The equation (1.5.14) is equivalent to equation (1.5.4).

Another Compact Formula

Assuming that the rifle is zeroed in at $x_T = OT$.

We want to change the zeroing in at $x_D = OD$ when we know the bullet trajectory height $y_D = DP$ (at range x_D).

Employing (1.2.12), (1.2.14), (1.5.4) and (1.5.14), we find that the change in sight-scope angle can be calculated using the following equation

$$\Delta\alpha_{hT} \approx (-\frac{y_D}{x_D} + \frac{h_D}{x_D} - \frac{h_T}{x_T}) \cdot \frac{180}{\pi}. \tag{1.5.15}$$

The angle of sight-scope that zeroes the rifle at $x_D = OD$, is

$$\alpha_{hD} = \alpha_{hT} + \Delta\alpha_{hT} \tag{1.5.16}$$

where α_{hT} and $\Delta\alpha_{hT}$ are estimated using respectively (1.5.6) and (1.5.14), or (1.5.15).

Height of Sight-Scope

The height of sight-scope, h_D, that zeroes the firearm at $x_D = OD$, is calculated employing (1.5.6), and (1.5.14), i.e. solving the equation:

$$\alpha_{hT} + \Delta\alpha_{hT} = -\frac{h_D}{x_D} \cdot \frac{180}{\pi}. \tag{1.5.17}$$

Number of Clicks to Adjust the Scope-Sight Angle

Usually the sight adjustment is done when zeroing the rifle at a certain horizontal range.

Every telescopic sight scope shows the number of clicks the marksman has to dial to change the sight-scope angle by a quantity in MOA.

For an iron sight, the rate of change in sight clicks depends on the distance between the front sight and rear sight.

For example, telescopic sights usually are set up to change the sight-scope angle at a

$$rate = (1/4) \ MOA = (0.25)MOA \ \text{(for 1 click).}$$

Thus, the number of clicks, corresponding to the change (1.5.14), or (1.5.15) is

$$N\_clicks = \frac{(\Delta\alpha_{hT})}{rate}. \qquad\qquad (1.5.18)$$

Note that the Example 5.2 illustrates the correction of the angle of sight-scope.

Example 5.1 Aiming angle
A marksman fires a 0.30 ball M2 bullet with an initial speed of 2800 ft/s.
The drop of the bullet at the horizontal range $x_D = 100$ yards and $x_T = 500$ yards from the gun is respectively $\bar{y}_D = -2.80$ inches and $\bar{y}_D = -76.50$ inches.

(a) Where does the bullet hit a target board located at a distance of $x_D = 100$ yards from the gun if the firearm is zeroed at $x_T = 500$ yards?

(b) Show as well, the aiming angle that zeroes in the rifle at $x_T = 500$ yards.
The sight-scope height that zeroes the rifle at 500 m is 1.5 inches.

Solution
(a) Substituting in (1.3.4), we find that the trajectory height at $x_D = 100$ yards is

$$y_D = x_D \left(\frac{|\bar{y}_T|}{x_T} - \frac{|\bar{y}_D|}{x_D} \right) = 100 \cdot \left(\frac{76.50}{500} - \frac{2.80}{100} \right) = 12.50 \ inches \ .$$

The point P that is $y_D = 12.50$ inches above the horizontal line is called **control point**. **Control Point P** controls the accuracy of zeroing the rifle at $x_T = 500$ yards, by measuring (at 100 yards) the vertical deviation of the group of shots from P.

(b) Substituting in (1.5.7) we find that the aiming angle that zeroes the rifle at $x_T = 500$ yards, and let the bullet pass $y_D = 12.50$ inches over the center of target located at $x_D = 100$ yards is

$$\alpha_{ST} \approx \left(\frac{|\bar{y}_T|}{x_T} + \frac{h_S}{x_T} \right) \cdot \frac{180}{\pi} = \left(\frac{76.5}{500 \cdot (36)} + \frac{1.5}{500 \cdot (36)} \right) \cdot \frac{180}{\pi} = 0.2483° \ .$$

The aiming angle in MOA is

$$\alpha_S = 0.2483 \cdot (60) = 14.898 MOA \ .$$

Example 5.2 Step by Step Aiming Adjustment
A rifle firing a 0.308 Sierra 165 grain Spitzer boattail bullet is zeroed at the range $x_T = OT = 250$ yards (scope height $h_T = 1.8$ inches). Departure line is OR (fig. 2).
According to Sierra[1], at the zero range $x_T = 250$ yards the bullet drop is $\bar{y}_T = -16.61$ inches, while at zero-range $x_D = 100$ yards the bullet drop is $\bar{y}_D = -2.37$ inches.

(a) Find the departure angle and the aiming angle that zeros in the rifle at range $x_{OT} = 250$.

(b) Calculate the change in aiming angle, when we change the zero-range from $x_{OT} = 250$ yards to $x_D = 100$ yards.
Rate 1 click = 1/4 MOA (rate =0.25 MOA).

Solution
(a) The departure angle, that zeroes the firearm at $x_{OT} = 250$ yards is

$$\alpha_{OT} = \frac{|\bar{y}_T|}{x_T} \cdot \frac{180}{\pi} = \frac{16.61}{(36) \cdot 250} \cdot \frac{180}{\pi} = 0.1057° .$$

The angle of sight-scope at $x_T = 250$ yards is

$$\alpha_{hT} = -\frac{h_T}{x_T} \cdot \frac{180}{\pi} = -\frac{1.80}{250 \cdot (36)} \cdot \frac{180}{\pi} = -0.0115° .$$

Aiming angle is

$$\alpha_{ST} = \alpha_{OT} - \alpha_{hT} = 0.1057 - (-0.0115) = 0.1172° .$$

At $x_D = 100$ yards, with this scope set up, the bullet will pass above the horizontal line.
Substituting in (1.3.4), we find that the "height" $y_D = DP$ of bullet "above" the horizontal line at $x_D = 100$ yards is

$$y_D = x_D \left(\frac{|\bar{y}_T|}{x_T} - \frac{|\bar{y}_D|}{x_D} \right) = 100 \cdot \left(\frac{16.61}{250} - \frac{2.37}{100} \right) = 4.274 in.$$

So, if the rifle is zeroed at $x_T = 250$ yards then the bullet at $x_D = 100$ yards will pass $y_D = 4.274$ inches above the horizontal line.

(b) Assume that with the same scope height setup, $h_T = 1.8"$, we aim at the point D at the range $x_D = 100$ yards (LOS = DS, fig. 1).

The departure angle that zeroes the firearm at $x_D = 100$ yards is

$$\alpha_{0D} = \frac{|\bar{y}_{DT}|}{x_{0D}} \cdot \frac{180}{\pi} = \frac{2.37}{(36) \cdot 100} \cdot \frac{180}{\pi} = 0.0377°.$$

The corresponding sight-scope angle is

$$\alpha_{hD} = -\frac{h_T}{x_D} \cdot \frac{180}{\pi} = -\frac{1.80}{100 \cdot (36)} \cdot \frac{180}{\pi} = -0.02865°.$$

The aiming angle is

$$\alpha_{SD} = \alpha_{OD} - \alpha_{hD} = 0.0377 - (-0.02865) = 0.06635°.$$

As result, the change in aiming angle is

$$\Delta\alpha_S = \alpha_{SD} - \alpha_{ST} = -0.06635 - (-0.1172) = -0.05083°.$$

Consider equation (1.5.4),
$$\Delta\alpha_S = \Delta\alpha_{0T} - \Delta\alpha_{hT},$$
where
$\Delta\alpha_S = -0.05083°$.

According to the equation (1.5.4), for a fixed left side equal to, we can change one or the other term on the right side.

We must keep constant the change in departure angle α_{OD} to hit the target at range $x_D = 100$ meters.
To compensate we have to "increase" $\Delta\alpha_S$ to the value

$$\Delta\alpha_S = -0.05083° = -3.050 MOA.$$

That means that to the negative angle $\alpha_{hT} = -0.0115°$ (clock-wise rotation) we must add the above negative change in the angle of sight-scope.

The number of clicks, needed to correct the aiming angle, must be dialed in the direction that increases the height of sight scope.

$$N\_clicks = \frac{|\Delta\alpha_S|}{rate} = (\frac{3.050}{1/4}) = 12.20 \approx 12 \ clicks \ .$$

Employing (1.5.16) we find that the angle of sight-scope is

$$\alpha_{hD} = \alpha_{hT} + \Delta\alpha_{hT} =$$
$$= -0.0115° - 0.05085° = -0.06235° = -3.741 MOA.$$

The new angle is negative since it is obtained by a clock-wise rotation.

In absolute value, the new angle is bigger than the absolute value of the angle of sight-scope $\alpha_{hT} = -0.0115°$.

So, in order that the departure angle to remain unchanged, equal to $\alpha_{0D} = 0.0377°$, we have to increase the sight-scope angle by the quantity, in absolute value, $\Delta\alpha_h = |-3.050| = 3.050 MOA$.

The new sight scope height can be found using equation (1.5.17). Thus, we can write

$$-\frac{h_D}{x_D} \cdot \frac{180}{\pi} = -0.06235° \ .$$

Hence, the height of the adjusted sight scope is $h_D = 3.918$ inches.

Note. Sierra gives a change in angle of sight-scope of $\Delta\alpha_S = 3.05 MOA$, when the zero range changes from $x_T = 250$ yards to $x_D = 100$ yards.

Compact Solution Formula (1.5.14)

The change in angle of sight can be found using the compact formula (1.5.14). Indeed, we have:

$$\Delta\alpha_h \approx [\frac{|\bar{y}_D|}{x_D} - \frac{|\bar{y}_T|}{x_T} + \frac{h}{x_D} - \frac{h}{x_T}] \cdot \frac{180}{\pi} =$$

$$= [\frac{2.37}{100 \cdot (36)} - \frac{16.61}{250 \cdot (36)} + \frac{1.8}{100 \cdot (36)} - \frac{1.8}{250 \cdot (36)}] \cdot \frac{180}{\pi} = -0.0508°.$$

Employing Compact Formula (1.5.15)

We obtain the same result using the compact formula (1.5.15). Indeed,

$$\Delta\alpha_h = (-\frac{y_D}{x_D} + \frac{h}{x_D} - \frac{h}{x_T}) \cdot \frac{180}{\pi} =$$

$$= [\frac{-4.274}{100 \cdot (36)} + \frac{1.8}{100 \cdot (36)} - \frac{1.8}{250 \cdot (36)}] \cdot \frac{180}{\pi} = -0.05083°.$$

Example 5.3 Number of clicks to change the zeroing in.
A rifle firing a 0.308 Sierra 165 grain Spitzer boattail bullet is zeroed in at the range $x_T = 100$ yards.
The height of sight-scope above the bore line is $y_S = 1.8$ *inches* .
Rate: 1 click = 1/4 MOA (rate =0.25 MOA).

(a) For the range $x_D = 250$ yards, find the "height" of the bullet trajectory "above" horizontal line if the bullet drop at $x_D = 250$ yards is $\bar{y}_D = -16.61$ inches.
At zero-range $x_T = 100$ yards the bullet drop is $\bar{y}_T = -2.37$ inches.

(b) Change the zeroing from zero range $x_T = 100$ yards to $x_D = 250$ yards, and estimate the number of clicks needed to adjust the angle of sight-scope.

(c) Calculate the sight-scope height when rifle is zeroed in at $x_D = 250$ yards.

Solution
(a) Rifle is zeroed at $x_T = 100$ yards.
Substituting in (1.3.4), we find that the "height" of bullet "over" the horizontal line at $x_D = 250$ yards is

$$y_D = x_D(\frac{|\bar{y}_T|}{x_T} - \frac{|\bar{y}_D|}{x_D}) = 250 \cdot (\frac{2.37}{100} - \frac{16.61}{250}) = -10.685 in.$$

(b) At $x_D = 250$ yards the trajectory of bullet is $y_D = -10.685$ inches under the horizontal line.
Employing (1.5.15), we find that the change in sight-scope angle is

$$\Delta\alpha_{hT} = (-\frac{y_D}{x_D} + \frac{h_t}{x_D} - \frac{h_T}{x_T}) \cdot \frac{180}{\pi} =$$

$$= [-\frac{-10.685}{250 \cdot (36)} + \frac{1.8}{250 \cdot (36)} - \frac{1.8}{100 \cdot (36)}] \cdot \frac{180}{\pi} = 0.05083°.$$

The same result we obtain employing the compact formula (1.5.14).Indeed:

$$\Delta\alpha_{hT} \approx [\frac{|\bar{y}_D|}{x_D} - \frac{|\bar{y}_T|}{x_T} + \frac{h_T}{x_D} - \frac{h_T}{x_T}] \cdot \frac{180}{\pi} =$$

$$= [\frac{16.61}{250 \cdot (36)} - \frac{2.37}{100 \cdot (36)} + \frac{1.8}{250 \cdot (36)} - \frac{1.8}{100 \cdot (36)}] \cdot \frac{180}{\pi} = 0.05083°.$$

So, the change in the angle of sight-scope, when we change the zeroing at $x_D = 250$ yards is

$$\Delta\alpha_{hT} = 0.05083° = 3.050 MOA.$$

The number of clicks, needed to correct the aiming angle,

$$N\_clicks = \frac{|\Delta\alpha_{hT}|}{rate} = (\frac{3.050}{1/4}) = 12.20 \approx 12 \ clicks,$$

must be dialed in the direction that increases the sight scope height.

(c) The sight- scope angle at $x_T = 100$ yards, is

$$\alpha_{hT} = -\frac{h_T}{x_T} \cdot \frac{180}{\pi} = -\frac{1.8}{100 \cdot (36)} \cdot \frac{180}{\pi} = -0.02865°$$

The sight scope angle at $x_D = 250$ is

$$\alpha_{hD} = \alpha_{hT} + \Delta\alpha_{hT} =$$
$$= -0.02865° + 0.05083° = 0.02218°.$$

Substituting in equation (1.5.17):

$$\alpha_{hD} = -\frac{h_D}{x_D} \cdot \frac{180}{\pi},$$

we can write:

$$-\frac{h_D}{x_D} \cdot \frac{180}{\pi} = 0.02218°.$$

Substituting in the above equation $x_D = [250 \cdot (36)]$, and solving for h_D, we find that the height of the sight-scope at $x_D = 250$ yards is $h_D = -3.484$ inches.

The sight-scope height $h_D = -3.484$ inches is a negative number because it corresponds to a positive sight-scope angle (counter clockwise rotation, see (fig. 2).

The absolute value, $|h_D| = 3.484$ inches, represents the height of the sight scope at zero range $x_D = 250$ yards.

Comment
When we change the zeroing in one direction (example 5.2), or in the opposite direction (example 5.3), the change in angle (in absolute value) of the sight-scope is the same, $\Delta\alpha_{hT} = 0.05083°$.

As we can see from example 5.2 and example 5.3, the sight-scope angle, respectively at $x_D = 100$ yards and $x_D = 250$ yards, is not the same ($\alpha_{hD} = -0.06623°$ and $h_D = 3.918$ inches against $\alpha_{hD} = 0.02218°$ and $h_D = 3.484$ inches).

The discrepancies are related to the fact that in the range tables, for aiming angle calculations, there is given the height of trajectory, or the trajectory path, based on an arbitrary chosen height of sight-scope; (see table 11.1 and table 11.2, section 1.11, the sight height in meters).

Thus, the marksman set up his/her sight-scope at a given zero range, let's say 600 meters, using his/her own sight scope height that might be different from 0.04 meters.

1.6 Testing and Adjusting Aiming Angle

In practice of firing with small arms, it is necessary to test the sight-scope setup of the firearm to assure shooting accuracy.

Testing of the sight scope is done using firing tests. The firing tests can be made, for example:

- After repairing or adjusting the iron sight scope, parts of a rifle, etc.
- To assure that the sight scope is set up correctly to zero the firearm at the desired range.
- When a competitive marksman fires in atmospheric conditions that are not standard.
  For example, in a new location that is different from the location the marksman usually fires.
- In general, when the long range shooting is not made in standard atmospheric conditions.
- When the long range shooting data are obtained solving the system (5.6.1) of differential equation using the G_1, or G_7 drag function and a fixed ballistic coefficient.

Inaccuracy of Bullet Drop that is Predicted Using Reference G_1, or G_7 functions and a Fixed BC.

Use of reference G-functions, G_1, or G_7, to predict the drop of a bullet in long range shooting, might result in relatively large errors.
Thus, for example, according to Wikipedia[2], at a firing range of 1,200 meters, the drop of 0.338 GB528 19.44g Lapua bullet (initial velocity 830m/s), predicted using G_1-drag function and BC = 0.785, is 16.073 meters.

The drop of the same bullet at the same range, predicted using Doppler radar measurements is 16.571.

So, theoretically there is a change of -0.50 meters (16.073 - 16.571) between the respective drops. It means that the bullet will impact in vertical direction 0.50 meters above the center of the target.

The following example illustrates the sight-scope testing and correction method.

Example 6.1 Testing procedure
Consider the standard atmosphere TSA (Traditional Standard Atmosphere).
A marksman with a Russian rifle Simonov SKS fires some 7.62mm bullets with a departure speed of 735m/s on a control board located $x_D = 100$ meters from the rifle.
Using the standard range table of SKS rifle the shooter finds that the departure angle that zeroes the rifle at $x_T = 300$ meters is $\alpha_{0T} = 0.216°$, while the departure angle that zeroes the rifle at $x_D = 100$ meters is $\alpha_{0D} = 0.084°$.

The rifle has a telescopic sight. The angle of sight-scope is set up to zero the rifle at the horizontal range $x_T = 300$ meters when the shooting conditions are standard.
The scope height $h_T = 0.0381$ meter corresponds to zero range $x_T = 300$ meters.

After shooting with some bullets, the center of the group of shots at $x_D = 100$ meters is $DQ = 0.250$ meters above the lower edge of a rectangular board (aiming point).

(a) How high "above" the **control point P** of the board target does the rifle shoot with the actual sight-scope set up at $x_T = 300$ meters?
(b) Is the sight scope of the rifle set up correctly to be zeroed at $x_T = 300$ meters?
(c) Adjust the sight height to zero the firearm at $x_T = 300$.

Solution

The y-coordinate, $y_D = DP$, of the center of the group of shots at $x_D = OD$ (at 100 meters) in standard firing conditions must be

$$y_D = x_D(\alpha_{0T} - \alpha_{OD}) \cdot \frac{\pi}{180} = 100 \cdot (0.216 - 0.084) \cdot \frac{\pi}{180} = 0.230m \quad (1.61)$$

over the center of the target, and not $DQ = 0.250m$.

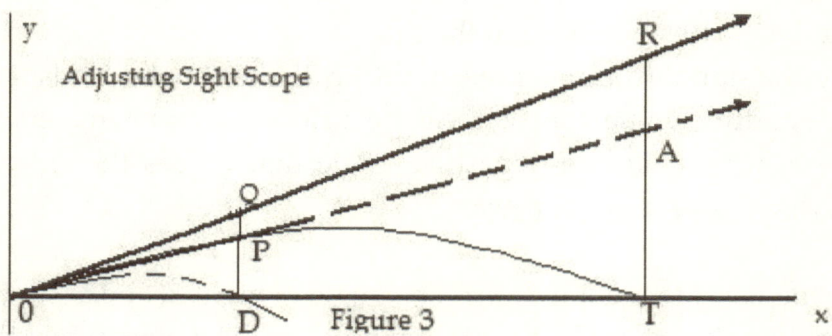

Figure 3

Control Point
The point P that is $y_D = 0.230$ meters above the horizontal line is called **control point**.
If the center of the group of shots, at range 100 meters, is at the point P then the sight scope is set up correctly, i.e. it zeroes the firearm at $x_T = 300$ meters.
The bullet will hit the center of the target at 250 meters.
Because of the causes listed above, the actual hit at the control board can be above or below the control point P.

(b) The rifle at 100 meters hits

$$PQ = DQ - DP = 0.250 - 0.230 = 0.02m$$

above the **control point P**.

That means that at $x_T = 300$ meters the rifle fires higher than it is predicted. The sight-scope is not set up correctly to hit the center of the target at $x_T = 300$.

The bullet, at $x_T = 300$, will hit above the center of the target.

(c) **Adjusting Sight Scope Height**
Correction of the sight-scope can be done by firing tests, using trial and error procedure, or by changing the height of sight employing theoretical procedure as it is shown below.

Since the angle of sight-scope, for the firearm zeroed at $x_T = 300$ is unchanged, the change in angle of sight-scope is $\Delta\alpha_h = 0$. Using (1.5.4), we find that the change in aiming angle is equal to the change in departure angle:

$$\Delta\alpha_S = \Delta\alpha_{0T} . \qquad (1.6.2)$$

Figure 3 shows that the deviation of the departure angle from the range value given in standard range table is

$$\Delta\alpha_{0T} = (\frac{y_Q - y_D}{x_D}) \cdot \frac{180}{\pi} , \qquad (1.6.3)$$

where $y_Q = DQ$, $y_D = DP$, P is the control point.
Substituting in (1.6.3) we have:

$$\Delta\alpha_{0T} = (\frac{y_Q - y_D}{x_D}) \cdot \frac{180}{\pi} = \frac{0.250 - 0.230}{100} \cdot \frac{180}{\pi} = 0.01146° = 0.688MOA .$$

Thus,
$$\Delta\alpha_S = \Delta\alpha_{0T} = 0.688MOA . \qquad (1.6.4)$$

To keep departure angle unchanged, i.e. in our example $\alpha_{OT} = 0.216°$, we have to reduce the angle of sight-scope by the correction quantity

$$\Delta\alpha_h = \Delta\alpha_S$$

Hence, the correction (degree) in sight-scope is given by the compact formula:

$$\Delta\alpha_h = (\frac{y_Q - y_D}{x_D}) \cdot \frac{180}{\pi} . \qquad (1.6.5)$$

Substituting in (1.6.5), we find the correction in angle of sight scope

$$\Delta\alpha_h = 0.688 MOA. \qquad (1.6.6)$$

When $y_D = 0$, i.e. the control point is at the point with coordinates $(x_D, y_D = 0)$, using (1.6.3) we can write:

$$\Delta\alpha_D = (\frac{y_Q}{x_D}) \cdot \frac{180}{\pi} . \qquad (1.6.7)$$

The equation (1.6.5) yields the change in angle of sight-scope,

$$\Delta\alpha_h = (\frac{y_Q}{x_D}) \cdot \frac{180}{\pi} . \qquad (1.6.8)$$

when the control point is $(x_D, y_D = 0)$.

Applying Compact Equation to Correct Aiming Angle (sighting angle)

Assume that a marksman has set up the rifle scope to zero the firearm at x_D. To zero the firearm at x_D, the marksman has used the data obtained employing G_1-drag function and a fixed coefficient.

The marksman fires some bullets at the target located at range x_D.

The vertical deviation of the center of the group of shots, at x_D is PQ and not zero. The actual deviation of the departure angle from the departure angle that needed to zero the rifle at x_D is equal to the value estimated using (1.6.7):

$$\Delta\alpha_{0D} = \frac{QP}{x_D}\cdot\frac{180}{\pi} = (\frac{y_Q}{x_D})\cdot\frac{180}{\pi}. \qquad (1.6.9)$$

Thus, the change in sight-scope angle is estimated by the equation

$$\Delta\alpha_h = (\frac{y_Q}{x_D})\cdot\frac{180}{\pi}. \qquad (1.6.10)$$

Comment
The correction (1.6.5) of the sight-scope can be seen as result of the change in standard conditions of shooting (for example as result of changes in atmospheric conditions, or other factors).
So, the aiming correction calculated in (1.6.5) is necessary to adjust the actual zero range that is greater than $x_T = 300$ meters.

Example 6.2 Shooting in non-standard conditions
A competitive shooter fires some 0.338 Lapua GB528 Scenar 19.44g bullet with a departure speed of 830 m/s on a control table located $x_D = 100$ meters from the rifle.
Using the standard range table 3.1 section 5.3, the shooter finds that the departure angle that zeroes the rifle at $x_T = 500$ meters is $\alpha_{0T} = 0.242°$, while the departure angle that zeroes the rifle at $x_D = 100$ meters is $\alpha_{0D} = 0.0419°$.

(a) Determine the control point P on the firing board, i.e. the height of the bullet trajectory above the horizontal line at $x_D = 100$ meters.

(b) Using ballistic tables, the shooter sets up the aiming angle in order to zero the rifle at $x_T = 500$ meters.

To test the scope set up, the shooter fires at a control table at $x_D = 100$ meters.

The center of the group of shoots is 0.360 meters above the horizontal line.

Is the sight scope of the rifle set up correctly to be zeroed at $x_T = 500$ meters in the actual shooting site?

How high is the bullet trajectory at $x_T = 500$ in the actual conditions?

Adjust the sight height to zero the firearm at $x_T = 500$.

Solution
(a) Substituting in (1.3.4) we find the height of the bullet trajectory (at $x_D = 100$ meters):

$$y_D = x_D(\alpha_{0T} - \alpha_D) \cdot \frac{\pi}{180} = 100 \cdot (0.242 - 0.0419) \cdot \frac{\pi}{180} = 0.349m \ .$$

(b) The sight scope is not set up correctly.
The sight scope correction is

$$\Delta\alpha_h = \frac{y_Q - y_D}{x_D} \cdot \frac{180}{\pi} = (\frac{0.360 - 0.349}{100}) \cdot \frac{180}{\pi} = 0.00630° = 0.378MOA \ .$$

At $x_T = 500$ the bullet trajectory height is (fig. 3)

$$\Delta y_T = x_T \cdot \Delta \alpha_{0T} = x_T \cdot (\frac{y_Q - y_D}{x_D}) = 500 \cdot \frac{0.360 - 0.349}{100} = 0.055m$$

above the center of the target.

Example 6.3 Testing the Sight Setup
A marksman fires some 0.338 GB528 19.44g Lapua bullets (initial velocity 830m/s).
The marksman has zeroed in advance the rifle at $x_T = 1,200$ meters, using a set of trajectory heights (or trajectory paths) predicted by a PC program that uses the G_1 function of resistance with a ballistic coefficient BC = 0.785 (see section 1.11).
At the end of shooting, the marksman measures a vertical deviation of $y_Q = 0.50$ meters of the center of the group shots over the center of the target located at $x_T = 1,200$ meters.
At $x_T = 1,200$ meters, the PC predicted drop of projectile is $\bar{y}_T = -16.073$ meters.
The height of sight scope of the rifle is $h_t = 2.7" = 006858m$.

(a) Determine the correction in aiming angle needed to shoot the center of target at $x_T = 1,200$ meters.
(b) Determine the control point at range $x_D = 100$ meters.

Solution
The marksman has setup the sight scope to zero the rifle at $x_D = 1,200$ meters using a departure angle of

$$\alpha_T = \frac{|\bar{y}_T|}{x_{TD}} \cdot \frac{180}{\pi} = \frac{16.073}{1200} \cdot \frac{180}{\pi} = 0.7674°.$$

The corresponding aiming angle is

$$\alpha_{ST} \approx \alpha_T + (\frac{h_T}{x_T}) \cdot \frac{180}{\pi} = (0.7674) + (\frac{0.06858}{1200}) \cdot \frac{180}{\pi} = 0.84774° .$$

(a) Employing (1.6.3), with control point $y_P = 0$, we have a change of

$$\Delta\alpha_T = (\frac{y_Q}{x_T}) \cdot \frac{180}{\pi} = \frac{0.5}{1200} \cdot \frac{180}{\pi} = 0.02387°$$

in departure angle. The correct departure angle that really zeroes the rifle at $x_T = 1,200$ meters is

$$\alpha_{OT} = \alpha_T - \Delta\alpha_T = 0.7674 - 0.02387 = 0.7435° .$$

The corresponding aiming angle is

$$\alpha_S \approx \alpha_T + (\frac{h_T}{x_T}) \cdot \frac{180}{\pi} = (0.7435) + (\frac{0.06858}{1200}) \cdot \frac{180}{\pi} = 0.8238° .$$

Since we keep unchanged the real departure angle that zeroes the rifle at $x_D = 1,200$, we have to change the sight-scope angle with the quantity
$$\Delta\alpha_h = \alpha_{S2} - \alpha_{SD} = 0.8238 - 0.84774 = -0.02387° .$$

(b) The control point at 100 meters is

$$y_P = x_{100} \cdot \alpha_{0D} \cdot \frac{\pi}{180} = 100 \cdot (0.7435°) \cdot \frac{\pi}{180} = 1.30m$$

over the lower edge of the target board, located at 100 meters. Thus, at 100 meters the trajectory of the bullet must pass through the control point in order that the firearm should be zeroed at $x_T = 1,200$ meters.

1.7 Coordinate System Related with Line of Sight

In practice of shooting, the marksman is interested to know the **height of a flying bullet** (trajectory height) above the line of sight (LOS) when the rifle is zeroed at a certain range $x_T = OT$ (fig. 2). In other words, the marksman wants to know the height $y' = EP$ of the trajectory at any range $x = OT$.

(Some exterior ballistic authors refer to the height of trajectory above or below the line of sight as **bullet path**.)

Note that for the range beyond the zero-range $x = OT$, the distance of the bullet from the line of sight is negative.

Let's show a simple way to estimate the height of trajectory over the line of sight. For this we introduce a coordinate system that has the x-axis along the LOS.

In exterior ballistics of long range shooting with small firearms the origin of the system of coordinates is located at the upper point S of the rifle scope, the x'-axis is directed along the line of sight (ST), while the y'-axis is perpendicular to the LOS. That system of coordinates is denoted (x'Sy').

Actually, the origin of the coordinate system (xoy), fig. 2, section 1.2, used in exterior ballistics to predict the projectile trajectory is located at the muzzle of the firearm; the direction of x-axis is along the horizon, while the y-axis is perpendicular to it.

Since the trajectory elements are obtained from the system of differential equations (5.6.1) related with xoy system, we need to convert those data into the system x'Sy'.

Thus, for example, we need to convert the coordinates (x, y) of the trajectory projectile into the corresponding coordinates (x', y') relative to x'Sy'.

Let's find the relationship between those two coordinate systems (figure 2, section 1.2).

The system of coordinates x'Sy' is obtained from the system of coordinates xoy by shifting vertically x-axis and y-axis y_s units till the origin of coordinates will coincide with the point S and then rotating it clockwise with an angle equal to the angle of sight $\alpha_h = \angle 0TS$.

The relationship between the coordinates of a point P(x, y) on the trajectory and the coordinates (x', y') of the same point P, but in the x'Sy' system, when rotation is clock-wise, is:

$$x' = x\cos(\alpha_h) - y\sin(\alpha_h),$$
$$y' = x\sin(\alpha_h) + y\cos(\alpha_h) - y_s,$$

(1.7.1)

where
$y_s = h$ is the y-coordinate of point S. The height of the scope is h.

The coordinates (x, y) of a point on the trajectory when are known the coordinates (x', y') are:

$$x = x'\cos(\alpha_h) + y'\sin(\alpha_h),$$
$$y = -x'\sin(\alpha_h) + y'\cos(\alpha_h) + y_s.$$

(1.7.2)

The angle is positive, i.e. $\alpha_h > 0$.

Since the angle $\alpha_h = \angle 0TS$ is very small, it follows that $\cos\alpha_h \approx 1$ and $\sin\alpha_h \approx 0$. From (1.7.1) and (1.7.2), for the coordinates of the projectile, we have respectively:

$$x' = x,$$
$$y' = x\sin(\alpha_h) + y\cos(\alpha_h) - y_s,$$

(1.7.3)

and

$$x = x',$$
$$y = -x'\sin(\alpha_h) + y'\cos(\alpha_h) + y_S,$$ (1.7.4)

where $y_S = h > 0$.

Note that y' is the perpendicular distance of the point-mass bullet from LOS.

Illustration

(a) The coordinates of the projectile in (xoy) are respectively (500 m, 1.5 m), while the scope height is 1.5 inches (0.0381m). The sight angle is

$$\alpha_h \approx \frac{h}{x} \cdot \frac{180}{\pi} = \frac{0.0381}{500} \cdot \frac{180}{\pi} = 0.004366°.$$

Using (1.7.3) we find $x' = 500$, and

$$y' = 500 \cdot \sin(0.004366) + 1.5 \cdot \cos(0.004366) - 0.0381 = 1.5 \text{ m},$$

i.e. for the range 500 meters the height of the bullet over the line of sight is 1.5 meters.

(b) The coordinates of the projectile in (x'Sy') are respectively (500m, - 1,5m), while the scope height is 1.5 inches (0.0381m). Substituting in (1.7.4) we find:

$$x = 500$$

$$y = -500 \cdot \sin(0.004366) + (-1.5) \cdot \cos(0.004366) + 0.0381 = -1.5 \ m$$

For the range 500 meters the bullet trajectory is 1.5 meters below the line of sight.

Projectile Velocity and "LOS" Coordinate System

The projectile velocity v is not dependent on the system of coordinates, i.e. the projectile velocity, at any given point on the trajectory, is the same no matter which reference system of coordinates we use.

The above statement is true since from (1.7.3), or (1.7.4) it follows that the components of velocity along x-axis and y-axis are equal, i.e.

$$dx'/dt = dx/dt, \qquad dy'/dt = dy/dt.$$

The angle the velocity forms with the x-axis is different from the angle the velocity forms with the LOS, i.e. x'-axis (fig. 4).

Note that the line of sight Sx' is not a fixed coordinate axis. Its direction depends on the location of aiming point, for example T or D.
Thus, the point D, located on the ox axis that passes from the origin of coordinates (0, 0) through T, is below the LOS.

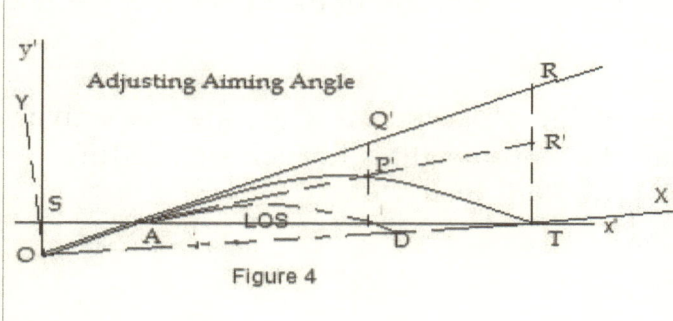

Figure 4

1.8 Height of Trajectory Over the Line of Sight

We assume that the firearm is zeroed at the horizontal range $x_T = OT$. The trajectory of bullet (fig. 2) passes at the point P which is above LOS, when the zeroing range ($x_T = OT$) is greater than $x_D = OD$, or below LOS, when the zeroing range, $x_T = OT$, is smaller than $x_D = OD$).

The height y'_P of the trajectory above LOS can be found using equation (1.7.1), i.e.

$$x'_D = x_D,$$
$$y'_D = x_D \sin(\alpha_h) + y_D \cos(\alpha_h) - h, \tag{1.8.1}$$

where

$$y_D = x_D(\tan\alpha_{0T} - \tan\alpha_D), \tag{1.8.2}$$

or

$$y_D = x_D(\frac{|\bar{y}_T|}{x_T} - \frac{|\bar{y}_D|}{x_D}), \tag{1.8.3}$$

where
α_{0T} is the departure angle that zeroes the firearm at the horizontal range, $x_T = OT$, α_D is the angle that zeroes the firearm at range $x_D = OD$, $\bar{y}_T$ and $\bar{y}_D$ are the bullet drops respectively at ranges $x_T = OT$ and $x_D = OD$, $y_S = h$ is the scope height.

When it is known the height y'_P of bullet trajectory above/below the line of sight, using (1.7.4), we find that the height y_D of the bullet above the horizontal line is

$$x_D = x'_D,$$
$$y_D = -x'_D \sin(\alpha_h) + y'_D \cos(\alpha_h) + h. \tag{1.8.4}$$

The following examples demonstrate the application of above formulae to find the height of trajectory above/below the line of sight, or the height of trajectory over/under the horizontal line.

Example 8.1 Lapua bullet, metric units

Refer to table 3.1, section 5.3.

Assume that the rifle is zeroed at $x_6 = 600m$. The departure angle that corresponds to the zero range $x_6 = 600m$ is $\alpha_{0T} = 0.3018°$.

Find the height y'_P of the bullet trajectory above/below the line of sight for the range $x_3 = 300m$ and $x_{10} = 1{,}000m$, if the sight height (scope above the bore line) is $h = 0.04m$.

The departure angle that corresponds to the zero-ranges: $x_3 = 300m$, $x_6 = 600m$ are respectively: $\alpha_3 = 0.1353°$ and $\alpha_{10} = 0.5958°$.

Solution

At $x_6 = 600$ meters, the angle of sight-scope, in absolute value, is

$$\alpha_h = \frac{h}{x} \cdot \frac{180}{\pi} = \frac{0.040}{600} \cdot \frac{180}{\pi} = 0.00382°.$$

(a) At $x_3 = 300m$, the height above the ground of the bullet trajectory (rifle zeroed at $x_6 = 600$ meters) is

$$y_D = x_D \cdot (\tan\alpha_0 - \tan\alpha_D) =$$

$$= 300 \cdot [\tan(0.3018) - \tan(0.1353)]) = 0.872m$$

Substituting in (1.8.1), we find that the height of bullet over the LOS is

$$y'_D = x_D \sin(\alpha_h) + y_D \cos(\alpha_h) - h =$$
$$= 300 \cdot \sin(0.00382) + 0.872 \cdot \cos(0.00382) - 0.040 = 0.852m.$$

This result is equal to the value 0.855 meters, for the given bullet, shown by Lapua[3]

(b) At $x_D = 1,000$ meters, the depth of the trajectory below horizontal line, is

$$y_D = 1,000 \cdot (\tan(0.3018) - \tan(0.5958)) = -5.1316m .$$

The distance of bullet trajectory from LOS is

$$y'_D = 1,000 \cdot \sin(0.00382) + (-5.1316) \cdot \cos(0.00382) - 0.040 = -5.105m.$$

The above result is quite equal to the value -5.082 shown by Lapua company[4].

Example 8.2 Lapua bullet, imperial units
The departure angle needed to hit the target at the zero-range equal to $x_T = 1,000$ *yards*, when the rifle fires a 0.338 Lapua GB528 Scenar 19.44g bullet, in ICAO atmosphere, is $\alpha_0 = 0.5247°$.
For the range $x_D = 800 yards$, find the height of the bullet trajectory above LOS if the scope above the bore line is $y_S = 1.6$ *inches* .
The departure angle that corresponds to the zero-range $x_D = 800$ *yards* is $\alpha_8 = 0.3880°$.

Solution
The sight-scope angle is

$$\alpha_h = \frac{h}{x} \cdot \frac{180}{\pi} = \frac{1.60}{1000 \cdot (36)} \cdot \frac{180}{\pi} = 0.00255°..$$

At range $x_D = 800$ yards, the height of the trajectory above the ground (ox axis) is

$$y_D = 36 \cdot x \cdot (\tan\alpha_{10} - \tan\alpha_8) = 36 \cdot 800 \cdot (\tan(0.5247) - \tan(0.3880)) = 68.7$$

For the range 800 yards, using the second formula of (8), we find that the height of the trajectory of bullet above the line of sight is

$$y'_D = 800 \cdot (36) \cdot \sin(0.00255) + 68.717 \cdot \cos(0.00255) - 1.60 = 68.40 in .$$

Example 8.3. Height of trajectory over horizontal line
The rifle firing a 0.338 Lapua GB528 Scenar 19.44g bullet is zeroed at the range 600 yards.
According to Lapua range table (see end note. 3), at range 1,000 yards, the height of the trajectory over the LOS is -158 inches. Find the height of the projectile below the horizontal line. The sight height is 1.6 inches.

Solution
The sight-scope angle is

$$\alpha_h = \frac{h}{x} \cdot \frac{180}{\pi} = \frac{1.60}{1000 \cdot (36)} \cdot \frac{180}{\pi} = 0.00255° .$$

Substituting in (1.8.4) we have:

$$x = x' = 1,000 \ .yards$$
$$y = -x'\sin(\alpha_h) + y'\cos(\alpha_h) + h =$$
$$= -1000 \cdot 36 \cdot \sin(0.00255) + (-158) \cdot \cos(0.00255) + 1.6 = -158 in.$$

1.9 Change of Zero Range Using Bullet Path

Assume that a firearm is zeroed in at a range $x_T = OT$ (figure 2, figure 4). The sight scope height is h. We need to adjust the aiming

angle in order to change the zeroing in at the new range $x_D = OD$
.The range $x_D = OD$ can be smaller or greater than $x_T = OT$.
Assume as well that we know the height $y'_D = DP'$ (fig. 2.) of the
trajectory above or below the horizontal line (bullet path).
The aiming angle at $x_D = OD$ is

$$\alpha_{SD} = \frac{y'_D}{x'_D} \cdot \frac{180}{\pi} . \qquad (1.9.1)$$

From figure 2, we can see that:

$$y'_D \approx x'_D (\alpha_{ST} - \alpha_{SD}) \cdot \frac{\pi}{180} , \qquad (1.9.2)$$

where
α_{ST} and α_{SD} are the aiming angles corresponding respectively to
$x'_D = OD$ and $x_T = OT$.
Since $x'_D = x_D$ and $y'_D = DP'$,we can write:

$$y'_D \approx x_D(\alpha_{ST} - \alpha_{SD}) \cdot \frac{\pi}{180} . \qquad (1.9.3)$$

Hence, for the change in aiming angle, when we change the
zeroing in from $x_T = OT$ to $x_D = OD$ we have:

$$\Delta\alpha_S = (\alpha_{SD} - \alpha_{ST}) = -\frac{y'_D}{x_D} \cdot \frac{180}{\pi} . \qquad (1.9.4)$$

The change in angle of sight-scope is

$$\Delta\alpha_h = \Delta\alpha_S . \qquad (1.9.5)$$

Substituting (1.9.4) into (1.9.5) we find that the change in angle of sight scope is

$$\Delta\alpha_h = -\frac{y'_D}{x_D} \cdot \frac{180}{\pi}. \qquad (1.9.6)$$

Example 9.1 Sight Adjustment
A rifle firing a 0.308 Sierra 165 grain Spitzer boattail bullet, with velocity 2,700 fps, is zeroed at the range $x_T = 250$ yards.
According to Sierra, at zero range $x_T = 250$ yards the bullet drop is -16.61 inches.

(a) Find angle of sight-scope.
(b) For the range 100 yards, find the height of the bullet trajectory above horizontal line.
(c) For the range 100 yards, find the height of bullet trajectory over LOS, if the height of sight scope above the bore line is $y_S = 1.8$ *inches* .
d) Find the sight adjustment when we change the zeroing from 250 yards into 100 yards.

Solution
(a) The sight-scope angle, in absolute value, at $x_T = 250$ yards is

$$\alpha_h = \frac{h}{x_T} \cdot \frac{180}{\pi} = \frac{1.80}{250 \cdot (36)} \cdot \frac{180}{\pi} = 0.01146°.$$

(b) Substituting in (1.8.3), we find that the height of bullet over the horizontal line at 100 yards is

$$y_D = x_D \left(\frac{|\bar{y}_T|}{x_T} - \frac{|\bar{y}_D|}{x_D} \right) = 100 \cdot \left(\frac{16.61}{250} - \frac{2.37}{100} \right) = 4.274 in.$$

(c) At 100 yards the height of bullet over LOS is

$$y'_D = 100 \cdot (36) \cdot \sin(0.01146) + 4.274 \cdot \cos(0.01146) - 1.80 = 3.194 in \,,$$

i.e. at 100 yards the bullet will hit 3.194 inches above the center of the target (aiming point), considering that the rifle is zeroed at $x_T = 250$ yards.
The corresponding change in angle of sight-scope is

$$\Delta\alpha_h = \frac{y'_D}{x_D} \cdot \frac{180}{\pi} = \frac{3.194}{100 \cdot (36)} \cdot \frac{180}{\pi} = 0.05073° = 3.05 MOA \,.$$

Example 9.2 Change of Zero Range
The departure angle needed to hit the target at the zero-range equal to $x_T = 600$ yards, when the rifle fires a 0.338 Lapua GB528 Scenar 19.44g bullet, in ICAO atmosphere, is $\alpha_{0T} = 0.2709°$ (velocity 2723.10 fps).
For the range $x_D = 900$ yards, find the height of the bullet trajectory below LOS if the scope above the bore line is $y_S = 1.6$ *inches* .
The departure angle that corresponds to the zero-range $x_D = 900$ yards is $\alpha_{0D} = 0.4535°$.

Solution
The angle of sight-scope at $x_T = 600$ yards is

$$\alpha_h = \frac{h}{x_T} \cdot \frac{180}{\pi} = \frac{1.60}{600 \cdot (36)} \cdot \frac{180}{\pi} = 0.00424° \,.$$

The height of bullet below the horizontal line at $x_D = 900$ yards is

$$y_D \approx x_D(\alpha_{0T} - \alpha_{0D}) \cdot \frac{\pi}{180} = 900 \cdot (36) \cdot (0.2709 - 0.4535) \cdot \frac{\pi}{180} = -103.26 in$$

At 900 yards the depth of bullet below LOS is

$$y'_D = 900 \cdot (36) \cdot \sin(0.00424) + (-103.26) \cdot \cos(0.00424) - 1.60 = -102.464i$$

The corresponding change in angle of sight-scope is

$$\Delta\alpha_h = \frac{y'_D}{x_D} \cdot \frac{180}{\pi} = \frac{-102.46}{900 \cdot (36)} \cdot \frac{180}{\pi} = -0.1812° = -10.87 MOA \ .$$

1.10 Change of Zero Range Using Trajectory Height

Assume that a firearm is zeroed in at a range $x_T = OT$. The sight scope height is h. We need to adjust the aiming angle in order to change the zeroing in at the new range $x_D = OD$.
The range $x_D = OD$ can be smaller or greater than $x_T = OT$.

Control point P corresponds to the shooting in standard conditions when the zero range is at $x_T = OT$ (figure 4).
Assume we know the bullet path $y'_D = DP'$ at range x_D when the zero range in standard conditions is $x_T = OT$.

As result of test shooting the center of group hits is $y'_Q = DQ'$.

The aiming angle that correspond to x_D, and to bullet path $y'_D = DP'$ is

$$\alpha_{SP'} = \frac{y'_D}{x_D} \cdot \frac{180}{\pi} \ . \qquad\qquad (1.10.1)$$

The aiming angle that correspond to x_D, and to bullet path $y'_Q = DQ'$ is

$$\alpha_{SQ} = \frac{y'_Q}{x_D} \cdot \frac{180}{\pi} . \qquad (1.10.2)$$

The change in aiming angle is

$$\Delta\alpha_S = (\frac{y'_Q - y'_{P'}}{x_D}) \cdot \frac{180}{\pi} . \qquad (1.10.3)$$

Thus, for the change in sight scope angle we have:

$$\Delta\alpha_h = (\frac{y'_Q - y'_{P'}}{x_D}) \cdot \frac{180}{\pi} . \qquad (1.10.4)$$

Let's simplify (1.10.3).
According to equation (1.8.1), we can express the bullet path through the height of trajectory over the horizontal line, i.e. we can write:

$$y'_P = x_D \sin(\alpha_h) + y_D \cos(\alpha_h) - h \qquad (1.10.5)$$

and

$$y'_Q = x_D \sin(\alpha_h) + y_Q \cos(\alpha_h) - h. \qquad (1.10.6)$$

Substituting in (1.10.3), we find that

$$\Delta\alpha_S = (\frac{y_Q - y_P}{x_D}) \cdot \cos\alpha_h \cdot \frac{180}{\pi} , \qquad (1.10.7)$$

where the sight-scope angle is

$$\alpha_h = -\frac{h}{x_T} \cdot \frac{180}{\pi} . \qquad (1.10.8)$$

Thus, we can write:

$$\Delta\alpha_h = (\frac{y_Q - y_D}{x_D}) \cdot \cos\alpha_h \cdot \frac{180}{\pi} , \qquad (1.10.9)$$

We see that there is an irrelevant difference between (1.10.9) and (1.10.4) since $\cos\alpha_h \approx 1$. We have to write (1.10.9) in the following form:

$$\Delta\alpha_h = (\frac{y_Q - y_D}{x_D}) \cdot \cos\alpha_h \cdot \frac{180}{\pi} . \qquad (1.10.10)$$

Example 10.1
A competitive shooter fires some 0.338 Lapua GB528 Scenar 19.44g bullet with a departure speed of 830 m/s on a control table located 100 meters from the rifle.

Using the standard range table 3.1, section 5.3, the shooter finds that the departure angle that zeroes the rifle at $x_T = 500$ meters is $\alpha_{0T} = 0.242°$, while the departure angle that zeroes the rifle at $x_D = 100$ meters is $\alpha_{0D} = 0.0419°$.
The control point P on the firing board, i.e. the height of the bullet trajectory above the horizontal line at $x_D = 100$ meters is 0.349.

Using ballistic standard tables, the shooter sets up the aiming angle in order to zero the rifle at $x_T = 500$ meters.
To test the scope set up, the shooter fires at a control board located at $x_D = 100$ meters.

The actual center of the group of shoots is 0.360 meters above the horizontal line. Adjust the sight height to zero the firearm at $x_T = 500$.

What is the height of the bullet trajectory at $x_T = 500$ at the actual conditions?

Solution
The sight scope is not set up correctly to zero the firearm at $x_T = 500$. The sight scope correction is

$$\Delta \alpha_h = -(\frac{y_Q - y_D}{x_D}) \cdot \cos \alpha_h \cdot \frac{180}{\pi} =$$

$$= -(\frac{0.360 - 0.349}{100}) \cdot \cos(0.00458) \cdot \frac{180}{\pi} = 0.0063°.$$

At $x_T = 500$ the bullet height is (see fig 4):

$$y_T = x_T \cdot \Delta \alpha_{0T} = x_T \cdot (\frac{y_Q - y_D}{x_D}) = 500 \cdot \frac{0.36 - 0.349}{100} = 0.055m$$

above the center of the target.

1.11 Trajectory Height and Bullet Path Tables

The standard range tables do not include the height of the trajectory above the line of sight (LOS), i.e. bullet path.
Using standard range tables and the equations obtained in the above sections, we will show the way we construct the Trajectory Path Tables.

For illustration we are going to construct the path table for the 0.338 Lapua GB528 Scenar bullet 19.44g (initial velocity 830 m/s, ICAO atmosphere), using the Standard Range Table 3.1 section 5.3. Scope height is $h = 0.040m$.

(a) **We assume that the rifle is zeroed at** $x_T = 600$ **meters** (departure angle $\alpha_{0T} = 0.3018°$).

Let's find the height of bullet trajectory above the LOS for ranges 100 meters till 700 meters.

Using (1.10.1), we find that the angle of scope-sight is

$$\alpha_h = \frac{h}{600} \cdot \frac{180}{\pi} = 0.0038° .$$

(1) Range $x_D = 100m$. Departure angle $\alpha_{0D} = 0.0419°$.

Using (1.8.2), we find the height of the trajectory over horizontal line:

$$y_D = x_D(\alpha_{0T} - \alpha_{0D}) \cdot \frac{\pi}{180} = 100 \cdot (0.3018 - 0.0419) \cdot \frac{\pi}{180} = 0.4536m .$$

The height of trajectory over LOS (the path) is:

$$y'_D = x_D \sin(\alpha_h) + y_D \cos(\alpha_h) - y_S =$$

$$= 100 \cdot \sin(0.0038) + 0.4536 \cdot \cos(0.0038) - 0.040 = 0.420m.$$

(2) Range $x_D = 200m$, departure angle $\alpha_{0D} = 0.0691°$.
We have:

$$y_D = x_D(\alpha_{0T} - \alpha_{0D}) \cdot \frac{\pi}{180} = 200 \cdot (0.3018 - 0.0691) \cdot \frac{\pi}{180} = 0.8123m .$$

$$y'_D = x_D \sin(\alpha_h) + y_D \cos(\alpha_h) - y_S =$$

$$= 200 \cdot \sin(0.0038) + 0.8123 \cdot \cos(0.0038) - 0.040 = 0.786m$$

(3) Range $x_D = 700m$, departure angle $\alpha_{0D} = 0.3662°$.
We have:

$$y_D = x_D(\alpha_{0T} - \alpha_{0D}) \cdot \frac{\pi}{180} = 700 \cdot (0.3018 - 0.3662) \cdot \frac{\pi}{180} = -0.7868m$$

$$y'_D = x_D \sin(\alpha_h) + y_D \cos(\alpha_h) - y_S =$$

$$= 700 \cdot \sin(0.0038) - 0.7868 \cdot \cos(0.0038) - 0.040 = -0.7801m$$

In the same way we find the height of bullet trajectory for other ranges.

(b) **We assume that the rifle is zeroed at** $x_T = 700$ **meters** (departure angle $\alpha_{0T} = 0.3662°$.). Let's find the height of bullet trajectory above the LOS for ranges 100 meters till 800 meters. Using (1.10.1), we find that the angle of scope-sight is

$$\alpha_h = \frac{h}{700} \cdot \frac{180}{\pi} = 0.00327°.$$

(4) Range $x_D = 600m$. Departure angle $\alpha_{0D} = 0.3018°$.
Using (1.8.2), we find the height of the trajectory over horizontal line:

$$y_D = x_D(\alpha_{0T} - \alpha_{0D}) \cdot \frac{\pi}{180} = 600 \cdot (0.3662 - 0.3018) \cdot \frac{\pi}{180} = 0.6744m.$$

The height (path) of trajectory over LOS is

$$y'_D = x_D \sin(\alpha_h) + y_D \cos(\alpha_h) - y_S =$$

$$= 600 \cdot \sin(0.00327) + 0.6744 \cdot \cos(0.00327) - 0.040 = 0.6687m.$$

Using the described procedure, we have constructed the following tables:

- Trajectory Height over the Horizontal Line, Table 11.1.
- Trajectory Height over LOS, Table 11.2.

Table 11.1 Trajectory Height: 0.338 Lapua Magnum Scenar GB528, 19.4g

Rifle Sighted	Trajectory Height Above Horizontal Line, [Meters]. Scope 0.040m						
	100	300	500	600	700	900	1000
100	0.000	-0.489	-1.746	-2.721			
200	0.048	-0.347	-1.508	-2.437			
300	0.163	0.000	-0.931	-1.744	-2.821		
400	0.252	0.267	-0.486	-1.210	-2.198		
500	0.349	0.559	0.000	-0.626	-1.517	-4.251	
600	0.454	0.872	0.522	0.000	-0.787	-3.31	
700	0.566	1.209	1.084	0.674	0.000	-2.300	
800	0.68	1.578	1.699	1.413	0.861	-1.19	-2.78
900	0.822	1.976	2.361	2.207	1.789	0.000	-1.452
1,000	0.967	2.413	3.087	3.079	2.805	1.307	0.000

Table 11.2   Trajectory Height. 0.338 Lapua Magnum Scenar GB528, 19.4g

Rifle Sighted	Trajectory Height Above LOS, [Meters]. Scope Height 0.040m						
	100	300	500	600	700	900	1000
100	0.000	-0.409	-1.586	-2.521			
200	0.027	-0.327	-1.449	-2.357			
300	0.136	0.000	-0.904	-1.704	-2.768		
400	0.222	0.257	0.476	-1.190	-2.168		
500	0.317	0.543	0.000	0.618	-1.501	-4.219	
600	0.420	0.852	0.515	0.000	-0.780	-3.291	
700	0.532	1.186	1.072	0.669	0.000	-2.290	
800	0.654	1.553	1.684	1.403	0.856	-1.19	-2.70
900	0.786	1.949	2.344	2.194	1.780	0.000	- 1.45
1,000	0.931	2.383	3.067	3.063	2.793	1.303	0.000

Both tables can be used to change the zeroing in, employing equations (1.5.15),

$$\Delta\alpha_{hT} = (-\frac{y_D}{x_D} + \frac{h_T}{x_D} - \frac{h_T}{x_T}) \cdot \frac{180}{\pi}$$

and equation (1.9.6)

$$\Delta\alpha_{hT} = -\frac{y'_D}{x_D} \cdot \frac{180}{\pi} .$$

The method 1 that uses equation (1.5.15) and table 11.1, can be used to change the zeroing in even when the height of sight scope of the rifle is not the sight-scope height predetermined to construct table 11.2 (unit meters in table 11.2).

The illustrations in example 11.1 and example 11.2, show the way we find the change in sight-scope angle.
The tables can be used to determine the Point-Blank Range, etc.

Example 11.1 Trajectory Height and Change of Sight
Use the data in table 11.1 to change the zeroing from $x_T = 600$ meters to $x_D = 500$ meters, when the height of the sight-scope is $h_T = 0.040$ meters.

Solution
Given that the rifle is zeroed at $x_T = 600$ meters, then at $x_D = 500$ meters the trajectory height over the horizontal line is $y_D = 0.522$ meters (table 11.1).
Employing (1.5.15), we find that the change in sight-scope angle is

$$\Delta\alpha_h = (-\frac{y_D}{x_D} + \frac{h_T}{x_D} - \frac{h_T}{x_T}) \cdot \frac{180}{\pi} =$$

$$= (-\frac{0.522}{500} + \frac{0.040}{500} - \frac{0.040}{600}) \cdot \frac{180}{\pi} = -0.05905°.$$

Example 11.2 Trajectory Height and Change of Sight
Use the data in table 11.1 to change the zeroing from $x_T = 600$ meters to $x_D = 500$ meters, when the height of the sight-scope is $h_T = 0.06858$ meters.

Solution
When rifle is zeroed at $x_T = 600$ meters, then at $x_D = 500$ meters the trajectory height over the horizontal line is $y_D = 0.522$ meters (table 11.1).
Employing (1.5.15), we have

$$\Delta\alpha_h = (-\frac{y_D}{x_D} + \frac{h_T}{x_D} - \frac{h_T}{x_T}) \cdot \frac{180}{\pi} =$$

$$= (-\frac{0.522}{500} + \frac{0.06858}{500} - \frac{0.6858}{600}) \cdot \frac{180}{\pi} = -0.05851.$$

Example 11.3 Bullet Path and Change of sight
Use the data of table 11.2 to change the zeroing in from 600 meters to 500 meters.

Solution
When rifle is zeroed at $x_T = 600m$, we find that the path of the trajectory over LOS at $x_D = 500$ meters is $y'_D = 0.515m$ table 11.2).

Substituting in (1.9.6), we find that the change in sight scope angle is

$$\Delta\alpha_h = -\frac{y'_D}{x_D} \cdot \frac{180}{\pi} = -\frac{0.515}{500} \cdot \frac{180}{\pi} = -0.05905° = -3.541 MOA .$$

Note that both methods, for the same sight scope height, give equal changes in sight-scope height.

Note
The results presented in table 11.1 and table 11.2 are compatible with the results presented by Lapua company at the Table .338 Lapua Magnum | Metric (page 3)
@Lapuaspecialpurpose2012Eng.pdf.

Thus, for example:
Lapua Table: Rifle sighted at 300 m, range 600 m, Impact Point below line of sight = -1711 mm (scope 40 mm above bore line).

Table 11.2 above: Rifle sighted at 300 m, range 600 m, Impact Point below line of sight = -1.704 m = - 1704mm (scope 40 mm above bore line)

2

Long Range Inclined Shooting

Introduction

In this chapter are shown some simple and practical methods to estimate the super elevation, or super depression angle in uphill, or downhill long-range shootings.
We employ the trajectory equation of projectile flying in vacuum to elaborate a prediction method of trajectories in Incline/declined shooting (section 2.7, 2.8).
The methods are approximate but incredibly accurate.

We also elaborate mathematically some of the exterior ballistic definitions, concepts, principles, formulas, etc., such as non-rigidity principle of the projectile trajectory.
We introduce **"The Non-Rigidity-Principle and Equal Drop"** model that can be used to find the super elevation, or super

depression angle and to set up the aiming angle in uphill, or downhill shooting.

We consider the point-mass projectile trajectory without taking into account the corrections for wind, rotation of bullet, rotation of Earth, etc.

In practice of long range shooting, to make feasible the process of shooting in the field of fast moving and hiding targets, the rifleman ignores many of the above factors that influence the accuracy of firing.
It is in the shooter discretion and in his/her shooting preparation to ignore or consider one or the other factor that influences the accuracy in long range shooting.

At the end of the chapter there are included some exercises that are solved using PC programs associated with the book EBLR. Through those exercises we verify the accuracy of approximate methods we use to predict the projectile trajectory, for example the approximate formulas related with inclined shooting.

2.1 A Practical Method on Inclined Shooting

In this section, and in the next section, we are introducing some simple equations (formulae) to estimate the inclined angle when we fire uphill or downhill in presence of drag.
They are based on the vacuum trajectory of projectile flying in absence of atmospheric air. (Section 2.7).

Super Elevation Angle
A simple and accurate approach to estimate the super elevation angle during uphill or downhill shooting is based on the following formula[5]:

$$\alpha_{OI} = \alpha_{OT} \cdot \cos(E), \qquad (2.1.1)$$

where

- $\alpha_{OT} = \angle TOT_1$ is the departure angle (in degree) that corresponds to the horizontal range $x_T = OT$ (in yards, or in meters),
- $E = R_x R$ is the elevation, or depression angle (in degree),
- $\alpha_{OI} = \angle ROR_1$ is the super elevation (super depression) angle that corresponds to the slant range $R = OR$, which is equal to the horizontal range $x_T = OT$, i.e. $R = x_T$ (fig. 5).

Note that the inclined range $R = OR$ is equal to the corresponding horizontal range, i.e. $R = x_T$. The horizontal range can be seen as an inclined shooting with elevation or depression angle E = 0.

Inclined Aiming Angle
The angle of sight α_S (in degree), for the inclined, or declined shooting, can be estimated using the following formula:

$$\alpha_S = \alpha_{OI} - \alpha_h, \qquad (2.1.2)$$

where
α_h is the sight-scope angle ($\alpha_h < 0$).
The sight-scope angle, that zeroes in the firearm at the inclined range $R = OR = x_T$ is

$$\alpha_h = -\frac{h}{R} \cdot \frac{180}{\pi} \qquad (2.1.3)$$

Since $R = x_T$, for the aiming angle we can write

$$\alpha_S = \alpha_{OI} + \frac{h}{x_T} \cdot \frac{180}{\pi}, \qquad (2.1.4)$$

where
h is the sight-scope height.

For the uphill shooting the elevation angle E is positive, while for the downhill shooting, the depression angle E is negative.

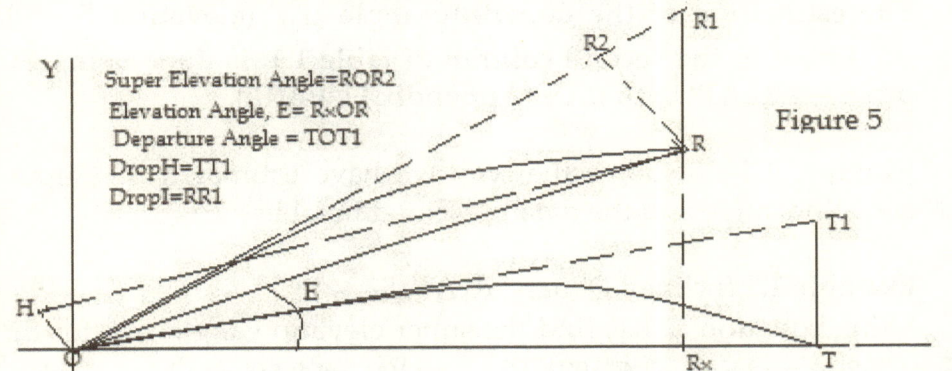

For Lapua Scenar GB528, 19.44g bullet, in ICAO atmosphere, the super elevation angle (estimated using equation (2.1.1)), for different inclined ranges, is given in table 1.1.

Table 1.1. Super Elevation Angle, 0.308 Lapua Scenar GB528, 19.44g Bullet, ICAO Atmosphere

Range	Elevation Angle (Degree)										
	0	15	20	25	30	35	40	45	50	55	60
(yards)	Super Elevation Angle (Degree)										
100	0.0379	0.0366	0.0356	0.0343	0.0328	0.0310	0.0290	0.0268	0.0244	0.0217	0.0190
200	0.0789	0.0762	0.0741	0.0715	0.0683	0.0646	0.0604	0.0558	0.0507	0.0453	0.0395
300	0.1218	0.1176	0.1145	0.1104	0.1055	0.0998	0.0933	0.0861	0.0783	0.0699	0.0609
400	0.1688	0.1630	0.1586	0.1530	0.1462	0.1383	0.1293	0.1194	0.1085	0.0968	0.0844
500	0.2178	0.2104	0.2047	0.1974	0.1886	0.1784	0.1668	0.1540	0.1400	0.1249	0.1089
600	0.2709	0.2617	0.2546	0.2455	0.2346	0.2219	0.2075	0.1916	0.1741	0.1554	0.1355
700	0.3273	0.3161	0.3076	0.2966	0.2835	0.2681	0.2507	0.2314	0.2104	0.1877	0.1637
800	0.3877	0.3745	0.3643	0.3514	0.3358	0.3176	0.2970	0.2741	0.2492	0.2224	0.1939
900	0.4535	0.4380	0.4262	0.4110	0.3927	0.3715	0.3474	0.3207	0.2915	0.2601	0.2268
1000	0.5247	0.5068	0.4931	0.4755	0.4544	0.4298	0.4019	0.3710	0.3373	0.3010	0.2624
1100	0.6005	0.5800	0.5643	0.5442	0.5200	0.4919	0.4600	0.4246	0.3860	0.3444	0.3003
1200	0.6831	0.6598	0.6419	0.6191	0.5916	0.5596	0.5233	0.4830	0.4391	0.3918	0.3416
1300	0.7731	0.7467	0.7264	0.7006	0.6695	0.6333	0.5922	0.5466	0.4969	0.4434	0.3865
1400	0.8698	0.8402	0.8173	0.7883	0.7533	0.7125	0.6663	0.6150	0.5591	0.4989	0.4349
1500	0.9758	0.9426	0.9170	0.8844	0.8451	0.7993	0.7475	0.6900	0.6272	0.5597	0.4879
1600	1.0913	1.0541	1.0255	0.9891	0.9451	0.8939	0.8360	0.7717	0.7015	0.6259	0.5457

In the second column of table 1.1, is given the departure angle that corresponds to a given horizontal range (100 - 1600 yards), when the elevation angle E is zero.

The estimation of the departure angle α_{0T} (elevation E = 0), presented in the second column of table 1.1, is done using PC program RLAPUA16.BAS (Appendix D, EBRL).

Example 1.1 illustrates the way we have estimated the super-elevation angle and the data presented in table 1.1.

Example 1.1 Inclined Departure Angle
Using equation (2.1.1) find the super elevation angle α_{0I} for the 0.338 Lapua Scenar GB528 19.44 g (300 gr) 8.59 mm bullet, when the inclined range and the elevation angle are respectively $R_S = 1200$ yards and $E = 35°$.
The departure angle for the horizontal range $x_T = 1200$ yards is $\alpha_{0T} = 0.6831°$ (see table 1.1, for E=0 degree.).

Solution
Substituting in formula (2.1.1) we find that the super elevation angle is

$$\alpha_{0I} = \alpha_{0T} \cdot \cos(E) = 0.6831 \cdot \cos(35) = 0.56°$$

Example 1.2
For 0.338 Lapua GB528 bullet, find the super depression angle needed to hit a target located on a declined plane 1200 yards from a sniper. The elevation angle is - 25 degree.
Again, in the first column of table 1.1, for the horizontal range 1200 yards, we find that .

Solution
Substituting in (2.1.1) we find that the super elevation angle is

$$\alpha_{OI} = \alpha_{OT} \cdot \cos(E) = 0.6831 \cdot \cos(-25) = 0.6191°.$$

Example 1.3 Change of Zeroing in Horizontal Shooting
0.388 Lapua GB528 19.44 bullet.
Assume that the rifle is zeroed at the horizontal range 100 yards.
Elevation angle E = 0. The sight height is $h = 2.7$ *inches* .

(a) Find the aiming angle α_S that zeroes the rifle at the horizontal range $R_S = 100$ yards.

(b) Find as well the sight angle α_S that zeroes the rifle at the horizontal range 1000 yards.

Solution
From table 1.1 (elevation angle $E = 0$) we find that the departure angle that corresponds to horizontal range 100 yards is .
In table 1.1 (elevation angle $E = 0$) we see that the departure angle that corresponds to horizontal range 1000 yards is .

(a) Horizontal range 100 yards
Substituting in (2.1.4) we find that the aiming angle is

$$\alpha_{S1} = \alpha_{OI} + \frac{h}{x_T} \cdot \frac{180}{\pi} = 0.0379 + \frac{2.7}{100 \cdot (36)} \cdot \frac{180}{\pi} = 0.08087°.$$

(b) Horizontal range 1000 yards: Substituting in (2.1.4) we find that the aiming angle is:

$$\alpha_{S2} = \alpha_{OI} + \frac{h}{x_T} \cdot \frac{180}{\pi} = 0.5247 + \frac{2.7}{1,000 \cdot (36)} \cdot \frac{180}{\pi} = 0.5290°.$$

Change in aiming angle is

$$\Delta\alpha_S = \alpha_{S2} - \alpha_{S1} = 0.5290° - 0.08087° = 0.4481° = 26.888 MOA.$$

Change in sight-scope angle is

$$\Delta\alpha_h = \Delta\alpha_S = 26.888 MOA.$$

Thus, we have to increase the sight scope angle with 26.888 MOA.

Number of Clicks
If we assume that 1 sight click is $1/8 = 0.125 \ MOA$, we find that the number of clicks to change zeroing in from 100 yards to 1000 yards is

$$\Delta N = 26.888/0.125 = 215.10 \approx 215 clicks.$$

Example 1.4 Change in Zeroing in Inclined Shooting
Find the inclined aiming angle α_S needed to zero the rifle respectively at the inclined range $R_{S1} = 100$ yards and $R_{S2} = 1,000$ yards if the elevation angle is $E = 30$ degree.

Solution
In table 1.1 (elevation angle $E = 30°$), for a range of 100 yards, we see that the super elevation angle is $\alpha_{0I} = 0.0328°$.
Substituting in (2.1.4), we find that the corresponding aiming angle is

$$\alpha_{S1} = \alpha_{0I} + \frac{h}{R_S} \cdot \frac{180}{\pi} = 0.0328 + \frac{2.7}{100 \cdot (36)} \cdot \frac{180}{\pi} = 0.07577°.$$

In table 1.1 (elevation angle $E = 30°$), for horizontal range 1,000 yards we see that the super elevation angle is $\alpha_{0I} = 0.4544$ degree. Using (2.1.4) we find that the aiming angle is

$$\alpha_{S2} = \alpha_{OI} + \frac{h}{R_S} \cdot \frac{180}{\pi} = 0.4544 + \frac{2.7}{1,000 \cdot (36)} \cdot \frac{180}{\pi} = 0.4587°.$$

Change in aiming angle is
$$\Delta\alpha_S = \alpha_{S2} - \alpha_{S1} = 0.4587° - 0.07577° = 0.3829° = 22.976 MOA.$$

Change in sight-scope angle is

$$\Delta\alpha_h = \Delta\alpha_S = 22.976 MOA.$$

Number of Sight-Scope Clicks

If we assume that 1 sight click is $1/8 = 0.125$ MOA, we find that the number of clicks to zero the rifle at the horizontal range 1000 yards is

$$\Delta N = 22.976/0.125 = 183.81 \approx 184 clicks.$$

Example 1.5 Change of Zeroing in Inclined Shooting
Use the results obtained in example 1.2 and example 1.3 to estimate the change in aiming angle and in sight clicks when we change the zeroing in.

(a) From the zero range 100 yards when the elevation angle is E = 0 to the zero range 100 yards when the elevation angle is E = 30 degree.
(b) From the zero range 1,000 yards when elevation angle is E = 0 to the zero range 1,000 yards when the elevation angle is E = 30 degree.

Solution
(a) Using the data in above examples we find that at 100 yards the change in aiming angle is

$$\Delta\alpha_S = \alpha_{S30} - \alpha_{S0} = 0.07577° - 0.08087° = -0.0051° = -0.306 MOA.$$

Change in angle of sight-scope is

$$\Delta\alpha_h = \Delta\alpha_S = -0.306 = -0.306 MOA.$$

The number of sight clicks is

$$\Delta N = 0.306/0.125 = 2.45 \approx 2clicks.$$

We have to reduce the sight scope clicks by 2 clicks.

(b) In the same way for 1,000 yards we find:

$$\Delta\alpha_S = \alpha_{S30} - \alpha_{S0} = 0.4587° - 0.5290° = -0.0703° = -4.21 MOA.$$

Change in angle of sight-scope is

$$\Delta\alpha_h = -\Delta\alpha_S = -(-4.21) = 4.21 MOA.$$

The number of sight clicks is

$$\Delta N = 0.421/0.125 = 33.74 \approx 34clicks.$$

We have to reduce the sight scope clicks by 34 clicks.

2.2 Basic Formula for Inclined Shooting

The approximate formula (2.1.1), is derived from the following equation[6] (see, as well, section 2.7 and 2.8)

$$\sin(2\alpha_{0I} + E) = \sin(2\alpha_{0T}) \cdot \cos^2(E) + \sin(E) \quad (2.2.1)$$

where

α_{0T} is the departure angle that corresponds to the horizontal range OT (elevation/depression) angle $E = 0$.

α_{0I} is the inclined angle that corresponds to the inclined range R. Formula (2.2.1) is determined using the trajectory equation of the projectile flying in absence of drag[7].

Hence, for the super elevation angle, we can write

$$\alpha_{0I} = \frac{\arcsin\left[\sin(2\alpha_{0T}) \cdot \cos^2(E) + \sin(E)\right] - E}{2} \quad (2.2.2)$$

Note

As it is shown in example 2.2, using the formula (2.2.2) we find that the super depression angle is slightly smaller than the super inclined angle.

That discrepancy is result of the term $\sin(E)$, which is positive for the uphill shooting and negative for the downhill shooting.

That result is normal since in uphill shooting the gravity acts "against" the "climbing" bullet while in declined shooting the gravity acts in "favor" of "descending bullet"

Example 2.1

Find the super elevation angle for the given Lapua GB528 19.44 bullet, if the inclined range is 1,000 yards and the inclined angle is 30 degree. In table 2.1, first column, we find that departing angle for horizontal range 1000 yards is $\alpha_0 = 0.5247°$.

Solution

Substituting in (2.2.2) $\alpha_0 = 0.5247°$, $E = 30°$, we find that the super elevation angle is:

$$\alpha_{OI} = \frac{\arcsin[\sin(2 \cdot 0.5247) \cdot \cos^2(30) + \sin(30)] - 30}{2} = 0.4565°$$

Using the approximate formula (2.1.1), we have found that the super elevation angle is $\alpha_{OI} = 0.4544°$ (see table 1.1).

Between the estimated elevation angles, calculated respectively using formula (2.1.1) and formula (2.2.2), there is an insignificant difference of

$$\Delta\alpha_{OI} = 0.4565 - 0.4544 = 0.0021° = 0.126 MOA,$$

or approximately 1 click.

Using (2.2.2), in the following table 2.1 there is estimated the departure angle for Lapua Scenar GB528, 19.44g bullet, fired in ICAO atmosphere with velocity 830 m/s = 2723.10 fps.

Table 2.1 Super Elevation Angle, Lapua Scenar GB528, 19.44g, ICAO

Range (yards)	Elevation Angle (Degree)								
	0	10	15	20	25	30	35	40	45
	Super Elevation Angle (Degree)								
100	0.0379	0.0366	0.0356	0.0356	0.0344	0.0328	0.0311	0.0290	0.0268
200	0.0789	0.0762	0.0742	0.0742	0.0715	0.0684	0.0647	0.0605	0.0558
300	0.1218	0.1177	0.1145	0.1145	0.1105	0.1056	0.0999	0.0934	0.0863
400	0.1688	0.1632	0.1588	0.1588	0.1532	0.1464	0.1385	0.1296	0.1196
500	0.2178	0.2106	0.2049	0.2049	0.1977	0.1890	0.1788	0.1673	0.1544
600	0.2709	0.2620	0.2550	0.2550	0.2460	0.2352	0.2225	0.2082	0.1922
700	0.3273	0.3166	0.3082	0.3082	0.2974	0.2843	0.2690	0.2517	0.2324
800	0.3877	0.3751	0.3652	0.3652	0.3524	0.3369	0.3188	0.2983	0.2755
900	0.4535	0.4389	0.4273	0.4273	0.4124	0.3943	0.3732	0.3492	0.3225
1000	0.5247	0.5080	0.4946	0.4946	0.4774	0.4565	0.4321	0.4043	0.3734
1100	0.6005	0.5816	0.5663	0.5663	0.5467	0.5228	0.4949	0.4631	0.4278
1200	0.6831	0.6619	0.6445	0.6445	0.6222	0.5951	0.5634	0.5273	0.4871
1300	0.7731	0.7493	0.7298	0.7298	0.7047	0.6741	0.6382	0.5974	0.5519
1400	0.8698	0.8435	0.8216	0.8216	0.7934	0.7590	0.7188	0.6729	0.6217
1500	0.9758	0.9467	0.9223	0.9223	0.8908	0.8524	0.8072	0.7558	0.6984
1600	1.0913	1.0593	1.0322	1.0322	0.9971	0.9542	0.9039	0.8464	0.7823

Atmosphere.

Note The method actually in use by some ballisticians, or shooters to predict the trajectory of bullets in uphill or downhill shooting, involves the estimation of the "Equivalent Horizontal Range"

(EHR). The EHR method is the basis to compensate for ballistic drop of the bullet [8].

The approximate method based on EHR is usually called the "rifleman's rule". Rifleman's rule is obtained for the projectile flying in the uniform gravitational force field and in absence of air resistance (see end note, ref.8).

The rifleman's rule is not so easy to be applied in practice of shooting, especially for long ranges. Moreover, for long ranges, the accuracy of estimation of the sighting angle is not satisfactory.

Downhill Shooting
The depression angle E for downhill shooting is negative i.e. $E < 0$.

In the same way as in the inclined shooting, using formulas (2.1.1) or (2.2.2), we can obtain similar tables as in the inclined shooting. As we will see in the following example, the declined angle of shooting is almost the same as the super-elevation angle in inclined shooting.

Example 2.2 Lapua GB528 19.44 bullet
For the declined range $R_S = 1000$ yards, the depression angle $E = -30$ degree and departure angle (table 2.1), using (2.2.2) we find that the declined angle of shooting is

$$\alpha_{OI} = \frac{\arcsin[\sin(2 \cdot 0.5247) \cdot \cos^2(-30) + \sin(-30)] - (-30)}{2} = 0.4523°.$$

Comparing the above value $\alpha_{OI} = 0.4523°$ of declined shooting with the corresponding value of inclined shooting $\alpha_{OI} = 0.4564°$

(table 2.1), we see that there is a difference, though relatively small.

2.3 **Non-Rigidity Principle of Trajectory**

Let's calculate the projectile drop $\bar{y}_{R1} = RR_1$ in inclined shooting (fig. 5). The angle $\angle R_2 RR_1$ is equal to the elevation angle E, i.e. $E = \angle R_2 RR_1$. Referring to the right triangle $R_2 RR_1$, we have:

$$RR_1 = \frac{RR2}{Cos(E)} \tag{2.3.1a}$$

From right triangle ROR2, we can write:

$$RR_2 = OR \cdot \tan(\alpha_{OI}) \approx x_T \cdot \alpha_{OI} \cdot \frac{\pi}{180} = x_T \cdot \alpha_{OT} \cdot \cos(E) \cdot \frac{\pi}{180}. \tag{2.3.2b}$$

Substituting (2.3.2b) into (2.3.1a), we find that the inclined drop of the projectile is equal to the horizontal drop, $\bar{y}_T = TT_1$, i.e.

$$RR_1 \approx x_T \cdot \alpha_{0T} \cdot \frac{\pi}{180} = y_T \tag{2.3.3c}$$

when

$$OR = x_T \text{ and } \alpha_{OI} = \alpha_{0T} \cdot \cos(E). \tag{2.3.4d}$$

Projectile Drop and Departure Angle
If we know the projectile drop, $\bar{y}_T = TT_1$, we can easily find the departure angle α_{0T} (Fig. 5).
Indeed, using triangle ORR_1, we can write:

$$\tan(\alpha_{0T}) = \frac{TT_1}{0T} = \frac{|\bar{y}_T|}{x_T} \tag{2.3.3}$$

Since α_{0T} is small, we can write approximately:

$$\alpha_{0T} \approx \frac{|\bar{y}_T|}{x_T} \cdot \frac{180}{\pi} \qquad (2.3.4)$$

Let's assume that the marksman fires on a target R that is uphill at a distance $OR = OT = x_T$. The elevation angle is E, while the drop RR_1 is equal to $\bar{y}_T = TT_1$, i.e. $RR_1 = \bar{y}_T = TT_1$.

The drop in uphill/downhill shooting and the distance to the target remain unchanged while the form of the trajectory changes. Thus,

$$RR_1 = \bar{y}_T = TT_1, \quad OR = OT = x_T \qquad (2.3.5)$$

Apparent Drop

In inclined shooting (fig. 5), the **apparent drop**, $A_{Drop} = RR_2$ is the vertical distance RR_2. We can write:

$$RR_2 = RR_1 \cdot \cos(E) \qquad (2.3.6)$$

Since $RR_1 = \bar{y}_T$ (in absolute value), we have:

$$|RR_2| = |\bar{y}_T| \cdot \cos(E) \qquad (2.3.7)$$

We can formulate the following statement that we call the **Non-Rigidity Principle of the trajectory.**

Non-Rigidity Principle

If we rotate the horizontal range $x_T = OT$ at an angle equal to the elevation angle E, then:

- Horizontal distance OT will be superimposed on inclined distance OR,

- Horizontal drop TT_1 will be superimposed on inclined drop RR_1,
- The form of projectile trajectory OKT is not identical to trajectory OLR, but changes in such a way that the **apparent drop** RR_2 decreases with increasing of elevation angle E (according to equation (2.3.6).

Based on the Non-Rigidity Principle we can estimate the apparent drop and the super elevation angle using respectively the equation (2.3.7) and the equation:

$$\alpha_{0I} \approx \frac{|RR_2|}{x_T} \cdot \frac{180}{\pi} \qquad (2.3.8)$$

The last equation is equivalent to the compact equation (2.3.1), i.e. to

$$\alpha_{0I} = \alpha_{0T} \cdot \cos(E).$$

Formulae (2.3.6) and (2.3.8) express the **Non-Rigidity Principle and Equal Drop Model** that allows us to find super elevation, or super depression angle, respectively in uphill or downhill shooting if we know the drop of projectile in horizontal shooting, or equivalently if we know the departure angle α_{0T} in horizontal shooting.

Example 3.1
A bullet is fired to hit a target located uphill, 1200 yards from the rifleman, at an elevation angle 25 degree.
Use the Non-rigidity principle to find the apparent drop, the super elevation angle and the departure angle, if the drop of the projectile at horizontal range 1,200 yards is 14.31 yards.

Solution
Apparent drop is

$$|RR_2| = |\bar{y}_T| \cdot \cos(E) = |-14.31| \cdot \cos(25) = 12.97 \; yards$$

The super elevation angle is

$$\alpha_{OI} \approx \frac{\overline{RR_2}}{x_T} \cdot \frac{180}{\pi} = \frac{12.97}{1200} \cdot \frac{180}{\pi} = 0.6193°.$$

The horizontal departure angle is

$$\alpha_{OT} \approx \frac{14.31}{1200} \cdot \frac{180}{\pi} = 0.683°$$

Exercise 3.2

A bullet is fired to hit a target located uphill, $OR = 1500$ meters from the rifleman, at an elevation angle $E = 30°$ .The drop of the bullet RR_1 is 18 meters.

Find super elevation angle, the horizontal departure angle and the respective drop (Fig. 5).

Solution

The horizontal departure angle is

$$\alpha_{OT} \approx \frac{TT_1}{oT} \cdot \frac{180}{\pi} = \frac{RR_1}{OR} \cdot \frac{180}{\pi} = \frac{18}{1500} \cdot \frac{180}{\pi} = 0.6875°.$$

The super elevation angle is

$$\alpha_{OI} \approx \alpha_{OT} \cdot \cos(E) = 0.6875 \cdot Cos(30) = 0.5954°.$$

2.4 Shooting with Departure Angle Zero

A bullet is fired horizontally with departure angle $\alpha_o = 0°$.

The bullet trajectory is under the x-axis. At the horizontal range $x_T = OT$ the bullet drop is $\bar{y}_{OT} = T_0 T$.

Suppose now that $\alpha_{0T} = \angle TOT_1$ is the departure angle that zeroes the firearm at the horizontal range $x_T = OT$. The bullet drop is $\bar{y}_T = TT_1$.

We rotate the trajectory "OATo" till the point "To" falls on x-

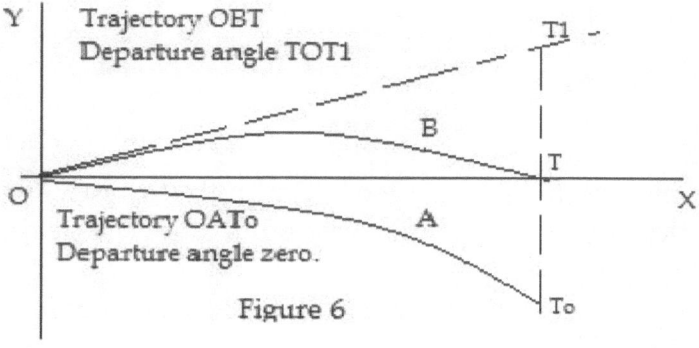

Figure 6

axis.

According to the Non-Rigidity of Trajectory model, the bullet drop is the same for both trajectories, i.e.

$$\bar{y}_T = \bar{y}_{OT} \tag{2.4.1}$$

So, for the departure angel angle α_{0T} , referring to the right triangle TOT₁, we can write:

$$\tan(\alpha_{0T}) = \frac{|\bar{y}_{OT}|}{x_T}, \tag{2.4.2}$$

or approximately

$$\alpha_{0T} \approx \frac{|\bar{y}_T|}{x_T} \cdot \frac{180}{\pi} \tag{2.4.3}$$

(Both drops are equal (equation (2.4.1)).

Thus,
If at a certain horizontal range $x_T = OT$, **we know the bullet
drop** $\bar{y}_{OT} = T_0T$ **when the bullet is fired with departure angle
zero** $\alpha_O = 0°$, **then the angle of departure,** α_{0T} , **needed to zero
the firearm at the horizontal range** $x_T = OT$ **is predicted using
equation (2.4.3).**

The departure angle in MOA (Minute of Angle) is

$$\alpha_{0T} \approx \frac{|\bar{y}_T|}{x_T} \cdot \frac{10800}{\pi}. \qquad (2.4.4)$$

Comment
In some range tables it is not given the departure angle but only
the bullet drop that corresponds to a given horizontal range.
The departure angle is unknown.
In such cases we can assume that the bullet is fired horizontally
(departure angle zero), or that the bullet is fired at whatever super
elevation angle.

For example, during experiments with Doppler radar to measure
the Mach number, the bullet is fired uphill at a given elevation
angle E with a given super elevation angle, that might be known
to us, or not.
Using the Non-rigidity trajectory model, we can find the
departure angle that zeroes the firearm at a given horizontal range
employing equation (2.4.3).

Example 4.1
A 0.338 Lapua GB528 bullet is fired horizontally. The projectile
drop at the horizontal range 1,500 meters is - 30.035 meters[9].

(a) Find the departure angle α_{0T} that zeroes the firearm at 1500 meters.

(b) Find the corresponding angle of sight, if the sight height h is $h = 2.7" = 2.7 \cdot 0.0254 = 0.06858m$.

(c) For the same inclined range (1,500m), find the super elevation angle if the elevation angle is E = 30 degree as well as the corresponding aiming angle.

(d) Find the super depression angle if E = - 30 degree.

Solution

Since the departure angle is not given at the Wikipedia paper, we can assume that the bullet is fired horizontally (departure angle zero), or, we can assume that the bullet is fired at whatever super elevation angle.

(a) Substituting in (2.4.3), we find that the departure angle that zeroes the firearm at horizontal range 1500 meters is

$$\alpha_{0T} \approx \frac{|y_T|}{x_T} \cdot \frac{180}{\pi} = \frac{30.035}{1500} \cdot \frac{180}{\pi} = 1.1473°.$$

(b) Substituting in (2.1.4), we find that the aiming angle is

$$\alpha_S = \alpha_{0T} + \frac{h}{R_S} \cdot \frac{180}{\pi} = 1.1473° + \frac{0.06858}{1500} \cdot \frac{180}{\pi} = 1.150° = 68.99 MOA.$$

(c) Employing (2.4.1), we find that the super elevation angle is

$$\alpha_{0I} = \alpha_{0T} \cdot \cos(E) = 1.1473 \cdot \cos(30) = 0.9936°.$$

(d) The super depression angle is

$$\alpha_{0I} = \alpha_{0T} \cdot \cos(E) = 1.1473 \cdot \cos(-30) = 0.9936°.$$

2.5 Equations of Trajectory and Accuracy of Non-Rigidity Principle

The super elevation, or super depression angle, as well as the other elements of the projectile trajectory can be obtained solving the differential equations of point-mass projectile trajectory[10]:

The general system of differential equations, that predicts the trajectory of a point-mass projectile flying in presence of drag and gravity acceleration, is [11]:

$$
\left\{
\begin{aligned}
\frac{dv_x}{dx} &= -\frac{\rho \cdot a}{\rho_0 \cdot a_0} \cdot c \cdot h(y) \cdot \frac{G_{D(v)}}{v} \\
\frac{dp}{dx} &= -\frac{g}{v_x^2} \\
\frac{dt}{dx} &= \frac{1}{v_x} \\
\frac{dy}{dx} &= p
\end{aligned}
\right\}
\tag{2.5.1}
$$

where

ρ and a are respectively the density of air and the speed of sound at the firing location, while ρ_0 and a_0 are respectively the density of air and the speed of sound at sea level (for ICAO standard atmosphere we denote ρ_{0N} and a_{0N} respectively density and speed of sound at the sea level).

The variable $p = \tan\alpha$, where α is the angle the projectile velocity forms with x-axis; (x, y) are the coordinates of projectile at a moment t, v is the projectile velocity (along the tangent to the trajectory); c is the ballistic coefficient (BC); $G_D(v)$ is the function of air resistance of the given projectile; $g = 9.80665 m / s^2$ (32.1740 ft/s²) is the standard value of the gravity acceleration

The component of projectile velocity along x-axis is $v_x = v \cdot \cos \alpha$, while $v_y = v \cdot \sin \alpha$ is the vertical component.

Projectile departure angle, at the initial point on trajectory ($x = 0$, y_0) is α_0, while the components of initial velocity are respectively $v_{x0} = v_0 \cos(\alpha_0)$, $v_{y0} = v_0 \sin(\alpha_0)$.

$h(y) = \rho/\rho_0$ is the density function; for example, it is given by the equation:

$$h(y) = \rho/\rho_0 = \rho_0 \cdot [(\tau_0 - 0.006328 \cdot y)/\tau_0)]^{4.4}.$$

where τ_0 is the virtual temperature at firing site; $\tau = (\tau_0 - 0.006328 \cdot y)$ is the virtual temperature at firing site (muzzle of firearm).
The density function $h(y)$ at the sea level is 1, i.e. $h(y) = 1$.
The system of differential equations of projectile trajectory, (2.5.1), can be written:

$$\begin{cases} \dfrac{dv_x}{dx} = -c \cdot \dfrac{p_0}{p_{oN}} \cdot \sqrt{\dfrac{\tau_{0N}}{\tau_0}} \cdot h(y) \cdot \dfrac{G_D(v)}{v} \\ \dfrac{dp}{dx} = -\dfrac{g}{v_x^2} \\ \dfrac{dt}{dx} = \dfrac{1}{v_x} \\ \dfrac{dy}{dx} = p \end{cases} \qquad (2.5.2)$$

where:

- P_{0N} and P_0 are the pressure of air respectively in standard atmosphere and at the firing site,
- τ_{0N} and τ_0 are the virtual temperature of air respectively in standard atmosphere and at the firing site.

To predict the projectile trajectory, the system of differential equations (2.5.1), or (2.5.2), can be solved numerically using the methods of numerical integration.

G-Function of Resistance, $G_D(v)$
Using the Doppler radar, or wind tunnels, there is found the drag coefficient as a function of Mach number, for a series of typical projectiles (see McCoy, Modern Exterior Ballistics, p.112). These functions, $C_D(M)$, are called **Reference Standard Drag Functions.**

In my books, we use the **Reference Standard G-function** of resistance $G_D(v/a_{0N})$ that is a function of projectile velocity, v.
In ICAO atmosphere, the G-function is related to Drag function by the equation

$$G_D(v/a_{0N}) = 4.81097 \times 10^{-4} v^2 \cdot C_D(v/a)$$

To compensate for missing dimensions of a point-mass projectile, it is introduced a correction factors, called Coefficient of Form, i. As result, in the system of differential equations, we have the BC written as:

$$c = i \cdot \frac{d^2}{m} \cdot 1000$$

Characteristic G-function of resistance
The estimation of the elements of a trajectory obtained by the system (2.5.2), using a reference G-function and BC of the projectile, is not that accurate, especially for relatively long ranges.

For that reason, for any given projectile (bullet), there is in use the corresponding Characteristic G-functions, obtained by Doppler radars.

The form coefficient, related with the characteristic G-function, is $i = 1$. The corresponding BC is

$$c = 1 \cdot \frac{d^2}{m} \cdot 1000$$

For example, the characteristic G-function of resistance of Bullet 0.338 **Lapua Scenar GB528 19.44 g (300 gr.)**, in ICAO atmosphere is

$$G_D(v) = 0.141 \cdot v - 30.039 , \qquad (2.5.3)$$

for $325 < v < 850$,

$$G_D(v) = 0.02438 \cdot v - 1.3696 ,$$

for $v \leq 325$

The characteristic G-function of 0.338 GB528 Lapua Scenar in ASM atmosphere is

$$G_D(v) = 0.139 \cdot v - 29.709 , \qquad (2.5.4)$$

for $330 < v < 850$

$$G_D(v) = 0.0242 \cdot v - 1.399 ,$$

for $v \leq 330$

More information is presented in "Exterior Ballistics: The remarkable Methods" by G. Klimi, p. 64 – 130, Xlibris, 2014.

Note. Since the G-function of resistance (2.5.2) is found using Doppler radar data, the horizontal range, obtained using the system of differential equations (2.5.2), is accurate.

Accuracy of the Non-Rigidity- Principle and Equal Drop Model
Solving the system of differential equations (2.5.1), or (2.5.2) for "Lapua Scenar GB528 19.44 g (300 gr.) bullet" using G-function of

resistance (2.5.4), there are found the elements of projectile trajectory, presented in table 5.1.

Table 5.1. Lapua Scenar GB528 19.44 g (300 gr.) bullet (Wikipedia)

Range (m)	0	300m	600m	900m	1,200m	1,500m
Velocity (m/s)	830	711	604	507	422	349
Time (s)	0	0.3918	0.8507	1.3937	2.0435	2.8276
Bullet Drop (m)	0	-0.715	-3.203	-8.146	-16.571	-30.035

Using the drop, given in Table 5.1, and the Non-Rigidity-Principle (formula 2.4.8), we have estimated the super elevation angle for elevation angle 5, 10, 15, 20, 30, 40 degree.

In table 5.2, it is added the super elevation angle for the range 100 meters, estimated using the drop 0.074 m shown in the last row of Table 5.2,second column[12].

Table 5.2

Range	100m	300m	600m	900m	1200m	1500m
Elevation	Super Elevation Angle [degree]					
0	0.0424	0.1366	0.3059	0.5186	0.7911	1.1471
5	0.0422	0.1360	0.3046	0.5162	0.7872	1.1407
10	0.0418	0.1344	0.3009	0.5099	0.7773	1.1258
20	0.0398	0.1282	0.2869	0.4858	0.7400	1.0706
30	0.0367	0.1181	0.2642	0.4971	0.6805	0.9836
40	0.0325	0.1044	0.2335	0.3950	0.6007	0.8676
Drop	-0.074	-0.715	-3.203	-8.146	-16.571	-30.035

In the following table 5.3, there is given the super elevation angle obtained solving numerically the system of equations (2.5.2).

Table 5.3. Numerical Integration. Lapua GB528 19.44.

Range	300m	600m	900m	1200m	1500m
Elevation	Super Elevation Angle [degree]				
0	0.1370	0.3061	0.5139	0.7920	1.1485
5	0.1360	0.3057	0.5117	0.7890	1.1420
10	0.1350	0.3015	0.5110	0.7764	1.1280
20	0.1285	0.2875	0.4876	0.7442	1.0684
30	0.1190	0.2700	0.4485	0.6825	0.9922
40	0.0328	0.2355	0.3972	0.6018	0.8750
Drop	-0.715	-3.203	-8.146	-16.571	-30.035

Comparing super elevation angle in table 5.2 and table 5.3, we see that "Non-Rigidity-Principle" of projectile trajectory gives incredibly accurate trajectory prediction.

Knowing the scope height, we can find the angle of sight-scope in MOA, for the uphill, or downhill shooting and then the numbers of sight clicks needed to adjust shooting.

We recommend to the shooter to prepare a card for the super elevation, or depression angle based on the shooting data for the horizontal range. Then using the Non-Rigidity-Drop model (Formulae (2.4.6), or (2.4.8)), we find the super elevation, or super depression angle.
In table 5.4, there is given the super-elevation angle for ranges in Imperial units (yard) obtain using formula (2.4.1).
We input to (2.4.1) the horizontal departure angle, for elevation or depression angle E = 0.

In general, comparing the data in table 5.1 and table 5.4, we can see that the approximate formula 2.1.1 (as well as formula (2.1.4) and equal-drop and non-rigidity model) can be always used for calculations.

Formula (2.1.1) is convenient for its practical simplicity.

Table 5.4. Lapua GB528 19.44g bullet. Non-Rigidity.
Imperial Units (yard).

Range	300	600	900	1,200	1,500
Elevation		Super Elevation Angle			
0	0.1218	0.2709	0.4535	0.6831	0.9758
5	0.1213	0.2698	0.4515	0.6798	0.9706
10	0.1199	0.2666	0.4460	0.6713	0.9581
20	0.1144	0.2542	0.4250	0.6393	0.9117
30	0.1055	0.2341	0.3912	0.5881	0.3880
40	0.0932	0.2069	0.3456	0.5193	0.7394

Summary
We have introduced some practical simple models that allow the shooters to set up the rifle to shoot uphill, or downhill.
Any long-range shooter, for training purposes, can prepare his or her own elevation shooting card based on the method presented in this section.
The shooter can as well prepare an app to find automatically the angle of sight in uphill or downhill shooting.
The long-range marksman can use his own range tables (for E = 0) that are obtained in the practice of field shootings, and based on that table, the marksman can prepare the uphill, or downhill shooting range table.

2.6 Introduction to Exterior Ballistics PC Programs

The reader can find the PC programs in the book "Elements of Exterior Ballistics: Long Range Shooting" (EBLR) published by Xlibris, 2016.

There are three main Exterior ballistics PC programs associated with the book. The PC programs (codes), in quick basic (QB), were compiled in 1991 by Col. Genc Kokoshi (ex-professor at the

Academy of General Staff, Tirana). The PC programs are continuously modified by the author, to reflect the advancements in Exterior Ballistics.

The three initial PC programs (prepared in 1991) were based on Siacci's G-function and were valid only for a type of artillery projectile, when field cannon shooting is done in TSA atmosphere. The core of all PC programs developed since 2005 are the code's that Col. Kokoshi has designed in 1991.

The QB codes designed by Col. Kokoshi are fascinating compact codes that made possible for me to advance in the study of exterior ballistics.

The Programmable TI58c, I was using in 80's, had no capacity to solve complex exterior ballistics problems though it was an amazing technology that helped me in my studies, solving few exterior ballistics problems related with the trajectory of projectile launched with low velocities, under 250 m/s.

The PC programs are compiled based on the system of differential equations (2.5.2) and G – function of resistance.

There are three types of PC programs.

1. The PC program that calculates the BC of a given projectile with respect to the reference G_1, G_7 or other reference G-function of resistance.
   The PC program named BC2016.Bas is shown in Appendix C (see (EBLR).

2. The PC programs that estimate the projectile range (and other elements of the trajectory) given the departure angle:

RPROJ16.BAS, RLAPUA16.BAS and RCHA16 shown in Appendix E.

- RPROJ16.BAS, predicts the elements of trajectory using the main reference G-functions, G_1, G_2, ...G_7, Siacci's and G_{43}-function (Russian, year 1943).

- RLAPUA16.BAS uses the **characteristic** G-function of GB528 Lapua Scenar bullet (Appendix B, G-functions B1 and B2).

- The PC program RCHA16 predicts the point of impact when it is known the departure angle. It is valid for bullets, presented in appendix B, and listed below.

3. The PC programs that calculates the departure angle needed to hit a given target located at a given point.

Such programs are:

- Alapua16.BAS, is shown in appendix D.
  program uses the **characteristic** G-function of 0.338 GB528 19.44, 8.59 mm, velocity 830m/s, BC = 3.796 (Appendixes B, B1 and B2).

- APROJ16.BAS, also shown in appendix D, calculates the departure angle using the main **reference** G-functions, G_1, G_2, ...G_7, Siacci's and Russian year 1943, G_{43}-function (Appendix A).

- The PC program ACHA16.BAS, shown in appendix D, is used to find the departure angle and other elements of the trajectory for the following list of bullets (Appendix B).

List of Bullets

For some bullets, listed below, there are constructed the characteristic G-functions of resistance that make possible to integrate numerically the system of differential equations of point mass projectiles (2.5.2), to predict the projectile trajectory with great accuracy.

1. Bullet 0.300 Winchester Magnum, Velocity 884m/s, BC = 4.97096:

2. Caliber 0.308, 168 grain, Sierra International Bullet, Velocity 792.48m/s, BC = 5.6325.

3. M118LR Bullet, 0.308" mass 0.01134 kg. Velocity 884m/s, BC = 5.3970.

4. M118 Ball Bullet (Federal GM308M2), 0.308", mass = 0.00134kg, velocity 792.48m/s, BC =5.3973

5. 300 gr., .338 - .416 Bullet (mass 0.01944kg, Velocity 927.40m/s, BC = 3.791

6. Caliber 0.30 Ball M2 Bullet, mass 0.00972 kg, Velocity 853.44 m/s, BC=6.2965

For each PC program there are given illustration examples to demonstrate the use of the program.

All three programs predict as well the projectile drop in any point when launching angle is zero.

Note on Downhill Shooting

The PC programs, Alapua16.bas, APROJ16.Bas, and ACHA16.Bas cannot predict the declined angle of shooting.

The following examples illustrate the solution of exterior ballistics problems using PC programs presented in the Appendixes of the book.

Example 6.1
(a) Find the ballistics coefficient c_7 related to the reference G_7-function for the 0.338 GB528 Lapua Scenar bullet launched in ICAO standard atmosphere with velocity $v_0 = 830 m/s$ if the point of impact is at the horizontal range $x_T = 1500m$ from the muzzle. There is no wind. The departure angle is 1.1471 degree.

(b) Find as well the BC for ranges 1200m, 900m, 600 m if the departure angle is respectively 0.7912-degree, 0.8186 degree, 0.3059 degree

(c) For the given bullet find the ballistics coefficient c_1 related to the reference G_1-function for the horizontal range $x_T = 1500m$.

Solution
    (a) Using BC2016.Bas (see the instructions in the PC program) we find: $c_7 = 3.6842 m^2 kg$.

We find as well impact velocity, 352.97 m/s and the time of flight, 2.8258 s.

In Imperial units BC is

$$C_7 = \frac{1.4222}{c_7} = \frac{1.4222}{3.6842} = 0.3860 lb/in^2 .$$

(b) Running the program with the given zero range and respective departure angle we find respectively:

Range 1200m: $c_7 = 3.690m^2 kg$, impact velocity 420m/s,

Range 900m: $c_7 = 3.721m^2 kg$, impact velocity 502m/s.

Range 600m: $c_7 = 3.789m^2 kg$, impact velocity 597m/s.

(c) The BC related with G_1-function is: $c_1 = 1.8683m^2 kg$.

Example 6.2
Find the ballistics coefficient c_7 related to the reference G_7-function for the 0.338 GB528 Lapua Scenar bullet launched in ICAO standard atmosphere with velocity $v_0 = 830m/s$ if the drop of the bullet at the horizontal range $x_T = 1200m$ is $\bar{y} = -16.571m$.

Solution
Use BC2016.Bas. Input the departure angle zero.

Note that when departure angle is zero, we have to input the y-coordinate of the muzzle of the firearm $\bar{y} = +16.571m$ over the ground, while the y- coordinate of target will be zero.
Running the program, we find: $c = 3.698m^2 kg$.

Example 6.3
Shooting with the 0.338 GB528 Lapua Scenar bullet in ICAO atmosphere.

(a) Find the departure angle α_{0T} needed to hit the target located at the zero range $x_T = 1800m$. Initial velocity is 830m/s. BC = 3.796 corresponds to characteristic Lapua G-function.

(b) Find the super elevation angle α_{0i} needed to hit the target located at the inclined range $D_T = 1800m$ if the elevation angle is $E = 30°$.

(c) For the same bullet, use the reference G_7 - function to find the departure angle needed to hit the target at horizontal zero range $x_T = 1800m$.

Consider ICAO atmosphere and the ballistics coefficient given by Wikipedia: $C = 0.377 \ lb/in^2$. BC in metric unit is $c = 1.4222/0.377 = 3.772m^2/kg.$

Solution
Using Alapua16.BAS.
(a) $\alpha_{0T} = 1.601$, $v = 311.85m/s$, $t = 3.747s$. $c = 3.796$

(b) Input the inclined range $D_T = 1800m$ and the elevation angle $E = 30°$:
Executing the PC program, we obtain the departure angle

$$\alpha_{0i} + E = 31.387°.$$

Since $E = 30°$, we find that super elevation angle is $\alpha_{0i} = 1.387°$.

Using the approximate formula (2.1.1), we find approximately the same value
$$\alpha_{OI} = \alpha_{0T} \cdot \cos(E) = 1.6069 \cdot \cos(30°) = 1.392°.$$

We can easily see that there is a difference of 0.28 minutes between two values of super elevation angle.

(c) We can use APROJ16.BAS.
Using APROJ16.BAS we find the departure angle $\alpha_{0T} = 1.649°$.

Comment

Comparing the approximate value $\alpha_{0T} = 1.649°$ with the exact value $\alpha_{0T} = 1.6069$ found in (a), we see that there is a difference of

$$\Delta\alpha_{0T} = 1.649 - 1.6069 = 0.0421° = 2.52'$$

Example 6.4

Shooting with M118 LR bullet (Long Range, sniper bullet), mass $m = 0.01134kg$, caliber $d = 0.0078232\ m$, departure velocity, $v_0 = 884\ m/s$, ballistic coefficient, $BC = 5.3970\ m^2/kg$ that corresponds to characteristic G-function B6, appendix B. Consider the ASM atmosphere.

(a) Find the departure angle α_{0T} needed to hit the target located at zero range $x_T = 1500m$.

(b). Find the super elevation angle α_{0i} needed to hit the target located at the inclined range $D_T = 1500m$ if the inclined plane is $E = 25°$.

(c) Solve case (b) when there is a tail-wind with velocity 5m/s.

Solution

Using PC program ACHA16.BAS.
Procedure:
Select ICAO = 2. Select M118LR = 5

Input:

X-coordinate of target = 1500. Y-coordinate of target = 0. Y-coordinate of firearm = 0. Departure velocity = 884. Temperature = 15. Temperature of propellant = 21.11. Pressure = 750. Humidity = 0.78. Projectile mass = 1 (any non-zero number if there is no change in mass). Change in mass = 0. Range-wind = 0. Cross wind = 0. BC = 5.3970.

Output:
Departure angle = 1.4992 degree. time = 3.332s. velocity 288.47m/s. Impact angle = -3.113 degree. Trajectory vertex has the coordinates (894m, 14.62m).

(a) Thus $\alpha_0 = 1.4992°$, $v = 288.49m/s$, $t = 3.331s$.

Note. The departure angle predicted by Raymond Von Wahlde and Dennis Metz is $\alpha_0 = 1.4929°$, while the velocity and time are respectively $v = 288.31m/s$ and $t = 3.339s$.

(b) Input the inclined distance $D = 1500m$ and the elevation angle $E = 25°$ together with other elements given in the exercise. Executing the PC program, we obtain the departure angle

$$\alpha_{0i} + E = 26.3516°.$$

Since $E = 25°$ we find that super elevation angle is $\alpha_{0i} = 1.3516°$. Using the approximate formula (2.1.1), we find the same value

$$\alpha_{OI} = \alpha_{0T} \cdot \cos(E) = 1.4929 \cdot \cos(25°) = 1.353°.$$

(c) Input the inclined distance $D = 1500m$, the elevation angle $E = 25°$, and the velocity of wind together with other elements given in the exercise.

Executing the PC program, we obtain the departure angle

$$\alpha_{0i} + E = 26.2173°.$$

Hence, we find that $\alpha_{0i} = 1.2173°$

Example 6.5

Shooting in ASM atmosphere with a **M118 Ball bullet** with velocity $v_0 = 792.48 \, m/s$. It is given the departure angle $\alpha_0 = 0.8993°$, (when elevation angle is zero).

Use the reference G_7 -function and the corresponding BC, $c = 5.9258 \, m^2/kg$.

Consider that in ASM atmosphere, the temperature of air, the temperature of propellant of propellant, the pressure and humidity are respectively 15 degree Celsius, 21.11 degree C., pressure 750mm Hg, humidity 78% = 0.78..

Find:

(a) The coordinates of the point on the trajectory that correspond to the horizontal distance x =1000 meter.

(b) The coordinates of the point on the trajectory that correspond to horizontal range x = 1000m if there is a constant tail range-wind of +4.5m/s.

(c) Find the point of impact at the inclined range 1000m, if the marksman fires with the same super elevation angle $\alpha_0 = 0.8993°$ when the elevation angle of shooting site is +25 degree. Weather is without wind.

(d) Find the answer to (c) if there is a range wind of +4.5 m/s

Solution

Using RPROJ16.BAS and the data given above. Input 1000 as horizontal range. Use ASM option (Press 1) and G7.

We find:

(a) Bullet will hit at the point with coordinates $x_T = 999.9m$ and $y_T = 0.31m$.

(b) $x_T = 998.97m$, and $y_T = 0.044m$.

(c) $x_T = 906.23m$, $y_T = 424.77m$ Inclined range 1000.84m. The bullet will pass 2.19m over the center of the target located at incline range 1000m.

(d) $x_T = 906.29m$, $y_T = 427.40m$ Inclined range 1002m. The bullet will pass 4.788m over the center of target located at the inclined range 1000m.

Example 6.6 Projectile Drop
A marksman fires a **M118 Ball bullet**. The shooting site is 100 meters over the sea level. The initial velocity of bullet is $v_0 = 792.48$ m/s, while BC of the given bullet with respect to the reference G_7 - function is $c = 5.9258$ m^2/kg .
Consider that, the temperature of air and of propellant are equal to 10 degree Celsius, pressure 750mm Hg, humidity 50%.

Find the drop of projectile that correspond to the horizontal distance x = 1200 meters.
(a) Select the ASM option (press 1).

(b) Select the ICAO option, (press 2).

Note that the y-coordinate of the muzzle is 100m.

Solution
We use RPROJ16.BAS and the data given above. Input 1200 as horizontal range.

Running the PC program, we find:
(a) The drop at 1200 meters is -26.78 meter.

(b) The drop at 1200 meters is -26.80 meter

The drop must be identical since the bullet is fired at the same shooting site and with the same data. Note that there is a small difference in the drop due to the numerical method of solution of differential equations.

Example 6.7 Downhill Shooting
Consider the example 2.2, section 2.2.
Use the PC program Rlapua16.bas to find the elements of the Lapua GB528 19.44 bullet. Declined range $R_S = 1000$ yards, the depression angle $E = -30$ degree. Consider the ICAO atmosphere and the initial velocity 830m/s and the super elevation angle $\alpha_{OI} = 0.4523°$, estimated in example 2.2 section 2.2.

Solution
Input:
ICAO atmosphere, GB528 Lapua, Inclined (Press 1), inclined distance $R_S = 1000 \cdot 0.9144 = 914.4m$, elevation (depression) angle $E = -30°$, super elevation angle $\alpha_{OI} = 0.4523°$, y-coordinate of muzzle $y_0 = R_s \cdot \sin(E) = 914.4 \cdot \sin(30) = 457.20m$, departure velocity 830m/s, temperature of air = 15 degree Celsius, temperature of cartridge = 21.11 degree Celsius, pressure 760mm Hg., humidity = 0, projectile mass = 1 (or whatever number different from zero), change in projectile mass = 0, range wind = 0, cross wind = 0, BC = 3.796.

Step size in this case should be 1, or any other number but not zero. The real value of the bullet mass is required to be input if we are studying the influence of small changes in mass of the bullet.

Output: Coordinates of impact point (791, 0.04). The impact angle is $\alpha_T = -30.605°$.

The bullet will hit 0.89 m before the center of the target located at the point with abscissa

$$x_T = R_s \cdot \cos(E) = 914.4 \cdot \cos(-30°) = 791.89m \text{ , and } y_T = 0.$$

The vertical deviation of the given bullet from the center of the target at the inclined range 914.4m is $\Delta y = -0.469m$, i.e. around 0.50 meters below the center.

2.7 Projectile Motion in Vacuum

Introduction

We present a non-traditional approach to study the trajectory of a point-mass projectile flying in vacuum in presence of the gravity only.

The mathematical model of vacuum trajectory is derived from the differential equations of point-mass projectile flight in presence of drag and gravity.

The projectile motion is restricted to projectiles thrown near the Earth's surface with relatively small speeds.

The gravity is considered to be constant, since the distance traveled and the maximum height of projectiles above the Earth are very small compared to the Earth's radius.

The resistance of air (drag) is ignored.

The vacuum trajectory has many applications in real life, and especially in Exterior Ballistics (for example see the applications presented in ref. [13] and ref. [14]). The results obtained using vacuum equations of the trajectory of flight, in many cases, give very good outcomes, approximate to the results obtained for the projectile flight in presence of drag and gravity.

Vacuum trajectory of the flying projectile gives incredibly accurate results to predict the super elevation/super depression angle of the projectile, flying in presence of drag, to hit a target at a slant range.

Note. This section is published at www.ResearchGate.net as a pre-print, by the author, in March 25, 2019.

Projectile Flight in Vacuum

For the point-mass projectile flying in vacuum, the density of air at firing sight is $\rho = 0$. Substituting $\rho = 0$ in the general equation (2.5.1), or (2.5.2) we obtain the system of differential equations of projectile trajectory in vacuum:

$$\begin{cases} \dfrac{dv_x}{dx} = 0 \\[2mm] \dfrac{dp}{dx} = -\dfrac{g}{v_x^2} \\[2mm] \dfrac{dt}{dx} = \dfrac{1}{v_x} \\[2mm] \dfrac{dy}{dx} = p \end{cases} \qquad (2.7.1)$$

Let's solve the system (2.7.1) to find the vacuum equations of the projectile trajectory, as well as the other elements.

Component of Velocity Along x-axis

Integrating the first equation of system (2.7.1), we obtain the x-component of velocity,

$$v_x = v_0 \cos(\alpha_0)$$

remain constant during the flight (equal to the initial x-component of velocity, $(v_x = v_{x0})$).

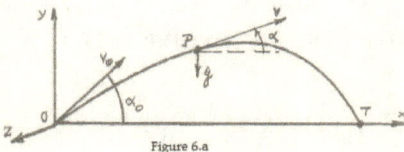

Figure 6.a

Equation of Projectile Trajectory

Substituting p, given in the fourth equation of system (2.7.1), into the second equation of (2.7.1) we have

$$\frac{d(dy/dx)}{dx} = -\frac{g}{v_x^2} \qquad (2.7.3)$$

Integrating the last equation, considering the initial conditions, as well as
$v_x = v_0 \cos(\alpha_0)$ and $p = \tan\alpha$, we obtain the equation of the point-mass projectile flying in vacuum

$$y = \tan(\alpha_0) \cdot x - \frac{g}{2v_0^2 \cos^2(\alpha_0)} \cdot x^2, \qquad (2.7.4)$$

where
$g = 9.0665 m/s^2$ is gravity acceleration.
Equation (2.7.4) describes a parabolic trajectory.

Angle of Flight
The second differential equation of system (2.7.1), after integration, yields the angle of flight α at any point on the parabola with coordinate (x, y):

$$\tan(\alpha) = \tan(\alpha_0) - \frac{g}{v_0^2 \cos^2(\alpha_0)} \cdot x \qquad (2.7.5)$$

The equation (2.7.5) can be obtained as well differentiating (with respect to x) the equation of the trajectory (2.7.4) and considering that $dy/dx = p$ and $p = \tan\alpha$.

Component of Velocity Along y-axis
Differentiating (2.7.4) with respect to time t, we find the y-component of projectile velocity at any point with abscissa x,

$$v_y = v_0 \sin(\alpha_0) - \frac{g}{v_0 \cos(\alpha_0)} \cdot x, \qquad (2.7.6)$$

Multiplying and dividing by dt the left side of the fourth equation of (2.7.1), i.e. $dy/dx = p$, we find another equation to calculate the vertical component of projectile velocity. Indeed, since

$$\frac{dy}{dt} \cdot \frac{dt}{dx} = p$$

and $v_x = dx/dt$, $v_y = dy/dt$ p $= \tan(\alpha)$, for the vertical component of the projectile velocity we can write:

$$v_y = \tan(\alpha) \cdot v_x,$$

where v_x and $\tan(\alpha)$ are calculated above.

The scalar-velocity of the projectile at any point with abscissa x
For the velocity of the projectile, we can write:

$$v^2 = v_x^2 + v_y^2 = v_0^2 - 2g \cdot tan(\alpha_0) \cdot x$$
$$+ \frac{g^2}{v_0^2 \cos^2(\alpha_0)} x^2 \qquad (2.7.7)$$

Parametric Equations of Projectile Trajectory

Integrating the third differential equation of system (2.7.1) we obtain the horizontal x-range as a function of time:

$$x = \cos(\alpha_0) \cdot t \qquad (2.7.8)$$

Substituting the time $t = x/\cos(\alpha_0)$ in (2.7.4), we find the trajectory equation as a function of time

$$y = \sin(\alpha_0) \cdot t - \frac{g}{2} \cdot t^2 \qquad (2.7.9)$$

Equations (2.7.8) and (2.7.9) are the parametric equations of projectile trajectory. They describe the coordinates of any point (x, y) as function of time.

Differentiating (2.7.9) with respect to t, we find the y-component of projectile velocity as a function of time t:

$$v_y = v_0 \sin(\alpha_0) - gt \qquad (2.7.10)$$

Projectile Velocity
For the scalar-velocity of projectile we find:

$$v^2 = v_x^2 + v_y^2 = v_0^2 - 2g[v_0 sin(\alpha_0) \cdot t] + g^2 t^2$$

Hence

$$v^2 = v_0^2 - 2gy + g^2 y^2 = (v_0 - gy)^2$$

and

$$v = v_0 - gy. \qquad (2.7.11)$$

System of Differential Equations, Variable (Time), t
The general system of differential equations (2.7.1), can be easily expressed as a system of differential equations of variable t .

Indeed, multiplying and dividing by dt each differential equation of (2.7.1), we have:

$$\begin{cases} \dfrac{dv_x}{dt} \cdot \dfrac{dt}{dx} = 0 \\[2mm] \dfrac{dp}{dt} \cdot \dfrac{dt}{dx} = -\dfrac{g}{v_x^2} \\[2mm] \dfrac{dt}{dx} = \dfrac{1}{v_x} \\[2mm] \dfrac{dy}{dt} \cdot \dfrac{dt}{dx} = p \end{cases} \qquad (2.712)$$

Hence

$$\begin{cases} \dfrac{dv_x}{dt} = 0 \\[2mm] \dfrac{dp}{dt} = -\dfrac{g}{v_x} \\[2mm] \dfrac{dx}{dt} = v_x \\[2mm] \dfrac{dy}{dt} = p v_x \end{cases} \qquad (2.7.13)$$

Integrating the differential equations (2.7.13), we find the elements of projectile trajectory related to variable time t.
The reader himself/herself can easily solve those differential

Note. If the initial position of the projectile, thrown with velocity $\vec{v}_0$ at an angle α_0 with the horizon, will be at the point with coordinates $(0, y_0)$, the equations of projectile trajectory are respectively:

$$y = y_0 + \tan(\alpha_0) \cdot x - \frac{g}{2 v_0^2 \cos^2(\alpha_0)} \cdot x^2, \qquad (2.7.15)$$

and

$$y = y_0 + \sin(\alpha_0) \cdot t - \frac{g}{2} \cdot t^2 \qquad (2.7.16)$$

Solution of (2.7.13) is:

$$
\rightarrow \quad \left\{
\begin{array}{c}
v_x = v_0 \cos(\alpha_0) \\[2mm]
\tan(\alpha) = \tan(\alpha_0) - \dfrac{g}{v_0 \cos(\alpha_0)} \cdot t \\[4mm]
x = v_0 \cos(\alpha_0) \cdot t \\[4mm]
y = \sin(\alpha_0) \cdot t - \dfrac{g}{2v_0^2} \cdot t^2
\end{array}
\right\} \qquad (2.7.14)
$$

When we fire a projectile, we aim to hit a target located at a known location (x, y) that can be at the same level as the firing point, above, or below it.

In other words, for a given location of the target, (x, y), and a known initial speed of the projectile v_0, we need to find that launching angle α_0, for which the projectile will hit the given target, the time of flight " t " to the target, the velocity v_T at the target, etc.

Now, let's find some important characteristics of vacuum projectile trajectory.

For that we consider the equation of the trajectory (2.7.4), i.e.

$$
y = (\tan\alpha_0)x - \left(\frac{g}{2v_0^2 \cos^2\alpha_0}\right)x^2
$$

Horizontal Range

At the point of impact that is at the same level as the gun, the y-coordinate is zero. Substituting $y=0$ in the trajectory equation, we have:

$$(\tan\alpha_0)x-(\frac{g}{2v_0^2\cos^2\alpha_0})x^2=0$$

Hence, we find

$$x=0 \text{ and } x_T=\frac{v_0^2}{g}\sin2\alpha_0. \qquad (2.7.17)$$

The first solution, $x=0$, corresponds to the launching point, while the second one corresponds to the impact point of projectile.

Launching Angle

From (2.7.17), for the launching angle needed to hit the target located at the horizontal range x_T we obtain:

$$\alpha_0=\frac{1}{2}\sin^{-1}(\frac{g\cdot x_T}{v_0^2}) \qquad (2.7.18)$$

Impact Speed

Substituting in (2.7.11), $y=0$ we find that the speed of the projectile at the point of impact x_T is the same as the speed v_0 of projectile at the launching point,

$$v_T^2=v_0^2-2gy=v_0^2-2g(0)=v_0^2 \qquad (2.7.19)$$

Impact Angle

Substituting the abscissa x_T of impact point into (2.7.4) we have:

$$\tan\alpha_T=\tan\alpha_0-(\frac{g}{v_0^2\cos^2\alpha_0})x_T=\tan\alpha_0-(\frac{g}{v_0^2\cos^2\alpha_0})(\frac{v_0^2}{g}\sin2\alpha_0)$$

Hence, since $\sin2\alpha_0=2\sin\alpha_0\cos\alpha_0$, we find that the impact angle in absolute value is equal to the launching angle, i.e.

$$\tan\alpha_T = -\tan\alpha_0 \text{ , or } \alpha_T = -\alpha_0 \qquad (2.7.20)$$

Time of Flight
Substituting $x_T = v_0^2 \sin2\alpha_0/g$ in the equation $x_T = (v_0\cos\alpha_0)t_T$, we find that the time of flight to the target is

$$t_T = \frac{2v_0\sin\alpha_0}{g} \qquad (2.7.21)$$

Trajectory Vertex
At the vertex of trajectory (m), the angle of flight is zero. We can write equation (2.7.5):

$$\tan(\alpha_m) = \tan(\alpha_0) - \frac{g}{v_0^2\cos^2(\alpha_0)} \cdot x_m$$

Substituting $\alpha_m = 0$, we have

$$\tan\alpha_0 - (\frac{g}{v_0^2\cos^2\alpha_0})x_m = 0$$

Solving for x_m the above equation, we find the abscissa of the vertex

$$x_m = \frac{v_0^2}{2g}\sin2\alpha_0 \qquad (2.7.22)$$

or, considering (2.7.17),

$$x_m = x_T/2 \qquad (2.7.23)$$

Substituting the above value in the equation of trajectory (2.7.4) we obtain the ordinate of the trajectory vertex that, for a given lunching angle, is the maximum altitude of the trajectory:

$$y_m = (\tan\alpha_0)x_m - (\frac{g}{2v_0^2\cos^2\alpha_0})x_m^2$$

$$= (\tan\alpha_0)(\frac{v_0^2}{2g}\sin2\alpha_0) - (\frac{g}{2v_0^2\cos^2\alpha_0})(\frac{v_0^2}{2g}\sin2\alpha_0) = \frac{v_0^2\sin^2(2\alpha_0)}{2g}$$

(2.7.24)

Thus, the trajectory vertex is located at the point with coordinates x_m and y_m determined respectively by (2.7.23) and (2.7.24).

Time of flight to the vertex point
Employing equation (2.7.22), and substituting in it the abscissa of the vertex we find the time of flight to the vertex

$$t_m = \frac{v_0\sin\alpha_0}{g}$$

(2.7.25)

or, considering (2.7.21), we find that

$$t_m = t_T/2$$

Maximum Range.
For a given projectile (given launching speed), the range (2.7.17),

$$x_T = \frac{v_0^2}{g}\sin2\alpha_0$$

depends on the launching angle " α_0 ".
The maximum value of sine function is 1. So, the range is maximum when $\sin2\alpha_0 = 1$, i.e. when the launching angle $\alpha_0 = 45°$. The value of the maximum range for $\alpha_0 = 45°$ is

$$x_{max} = \frac{v_0^2}{g}$$

(2.7.26)

Example 7.1.

A projectile is fired from a 60mm trench mortar with a speed of 89m/s under an angle of 45°. Determine the elements of the trajectory of flight.

Solution

Using the formulae obtained above, we find

The Horizontal Range

$$x_T = \frac{v_0^2}{g}\sin 2\alpha_0 = \frac{(89)^2}{9.80665}\sin(2 \cdot 45°) = 807.72m$$

Impact Speed

$$v_T = v_0 = 89m/s$$

Impact Angle

$$\alpha_T = \alpha_0 = 45°$$

Time of Flight

$$t_T = \frac{2v_0 \sin\alpha_0}{g} = \frac{2(89)\sin(45°)}{9.80665} = 12.83 \text{sec}$$

Coordinates of Trajectory Vertex

$$x_m = x_T/2 = 807.72/2 = 402.36m$$

$$y_m = \frac{v_0^2 \sin^2(2\alpha_0)}{2g} = \frac{(89)^2 \sin^2(90°)}{2(9.80665)} = 403.86m$$

Time of Flight to the Vertex Point

$$t_m = t_T/2 = 12.83/2 = 6.42 \text{sec}.$$

Table 7.1 compares some elements of the trajectory obtained above for the vacuum with the elements of trajectory of the same projectile in presence of resistance of air.

Table 7.1.

	Launching Angle	Angle of impact	Range (meter)	Time of flight, sec.	Maximum Altitude [m]
Ideal Trajectory	$\alpha_0=45°$	$\alpha=45°$	$x=807.72$	$t=12.84$	$y_m=403.86$
Drag Presence	$\alpha_0=45°$	$\alpha=48°10'$	$x=713.78$	$t=12.40$	$y_m=189$

From the table we see that the results obtained for the elements of the trajectory in free space are far from the results observed in practice.

Example 7.2.
An athlete throws a javelin of mass 0.80 kg. with initial speed $v_0=30m/s$. Launching angle is $\alpha_0=38°$. The launching point is $h=1.8m$ over the ground.
Find the distance of the impact point.

Solution
The launching point is located at the point $(0, 1.8)$. The equation of trajectory in this case is

$$y-y_0=(\tan\alpha_0)x-(\frac{g}{2v_0^2\cos^2\alpha_0})x^2$$

The impact point has the y-coordinate equal to zero, $y=0$. Substituting $y=0$, and $y_0=h=1.8$ we find that

$$\tan(38°)x-\frac{9.80665}{2\cdot30^2\cos^2(38°)}x^2+1.8=0$$

Solving for x the above second-degree equation we find that the horizontal range is $x=91.296m$.

2.8 Uphill and Downhill Shooting

We will show a simple formula to determine the super elevation angle α_{0I} for any vacuum trajectory, considering that the target is located at a point with coordinates (x, y), at a slant range $d = OR$ from the origin of coordinates (Figure. 5).
The angle of sight "α_{0I}" is the angle between the line of sight OR and axis of the bore.

We denote " E "the "Elevation Angle", i.e. the angle that the line of sight OR forms with x-axis. The Angle E is $-90° < E < 90°$.

Launching angle, α_0, included between x-axis and the initial velocity of projectile, is

$$\alpha_0 = E + \alpha_{0I} \qquad (2.8.1)$$

Consider the trajectory equation (2.7.4), i.e.

$$y = (\tan\alpha_0)x - \left(\frac{g}{2v_0^2 \cos^2 \alpha_0}\right)x^2 \qquad (2.8.2)$$

Substituting the coordinates of the target,

$$x = d \cdot \cos E, \qquad y = d \cdot \sin E, \qquad (2.8.3)$$

and the departure angle $\alpha_0 = E + \alpha_{0I}$, in the equation of the trajectory (2.8.2),
we can write

$$\sin(E) = \cos(E) \cdot \tan(E + \alpha_{0I}) - \frac{g \cdot \cos^2(E)}{2v_0^2 \cos^2(E + \alpha_{0I})} \cdot d \qquad (2.8.4)$$

Where $d = OR$ is the inclined distance to the impact point R.

Solving the last equation for the distance $d = OR$, we can write

$$d = \frac{2v_0^2 \cos^2(E+\alpha_{0I})}{g \cdot \cos^2(E)} \cdot \sin(E) \qquad (2.8.5)$$

The last equation is an equation that relates the slant range $d = OR$ and the super elevation angle α_{0I}, for a given initial velocity "v_0".

In practice, we are interested to determine the angle of sight "α_{0I}". Transforming the equation (2.8.5), we obtain a formula that allow us to find the angle "α_{0I}" as a function of the elevation angle "E", initial speed "v_0", and the slant distance "d", i.e.

$$\sin(2\alpha_{0I} + E) = \frac{gd}{v_0^2} \cos^2(E) + \sin(E) \qquad (2.8.6)$$

Let's rotate clock-wise the line of sight (d=OR), till it falls over x-axis. During the rotation, the range of inclined fire remains unchanged, d. As result, the elevation angle becomes zero, $E = 0$, the angle of sight α_{0I} changes (increases) and becomes α_{0T}, the slant range becomes horizontal range, ox_T.

Now, we apply formula (2.8.6) when the target is on the horizon, at the point $(0, x_T)$, at the same distance d from the launching point $x = 0$.
Substituting $E=0$ and $\alpha_{0I} = \alpha_{0T}$ in (4), we find

$$\sin(2 \cdot \alpha_{0T}) = \frac{gd}{v_0^2} \qquad (2.8.7)$$

This is the equation (2.7.17).
Substituting (2.8.7) in (2.8.6), we obtain the equation:

$$\sin(2\alpha_{0I} + E) = \sin(2\alpha_{0T}) \cdot \cos^2(E) + \sin(E), \qquad (2.8.8)$$

that we can apply to calculate the angle of sight α_{0I}.

Formula (2.8.8) relates the aiming angle "α_{0I}", that corresponds to the slant range "$d = OR$", with the launching angle (α_{0T}) that corresponds to the horizontal range "$d = ox_T$".

We can give to formula (2.8.8) a simple form when the angle of sight is small.

Employing the trigonometric identity for the sine of the sum of two angles, we can write

$$\sin(2\alpha_{0I}) \cdot \cos(E) + \cos(2\alpha_{0I}) \cdot \sin(E) =$$

$$= \sin(2\alpha_{0T}) \cdot \cos^2(E) + \sin(E) \qquad (2.8.10)$$

Now, we assume that

$$\cos(2\alpha_{0I}) \cong 1, \cos(\alpha_{0T}) \cong 1 \qquad (2.8.11)$$

i.e.

$$2\alpha_{0I} \approx 0, (\alpha_{0T}) \approx 0$$

Substituting in (2.8.10), we find a practical formula to estimate the angle of sight (super elevation angle/super depression angle)

$$\sin(2\alpha_{0I}) \approx \sin(2\alpha_{0T}) \cdot \cos(E) \qquad (2.8.12)$$

or

$$2\sin(\alpha_{0I})\cos(\alpha_{0I}) \approx 2\sin(\alpha_{0T})\cos(\alpha_{0T})\cos(E) \qquad (2.8.13)$$

From the above equation (since $\cos(\alpha_{0I}) \approx 1$, $\cos(\alpha_{0T}) \approx 1$), we obtain an approximate formula to estimate the angle of sight:

$$\sin(\alpha_{0I}) \approx \sin(\alpha_{0T}) \cdot \cos(E) \qquad (2.8.14)$$

for small angles, α_{0I} and α_{0T}. Since the angles are relatively small, from (2.8.14), we get

$$\alpha_{0I} \approx \alpha_{0T} \cdot \cos(E) \qquad (2.8.15)$$

Remarks.
Though the equations (2.8.8), (2.8.10), (2.8. 12), (2.8.14) and (2.8.15) are obtained for vacuum trajectory, they can be used also when projectile flies in standard atmosphere, in presence of drag, on condition that the angle of sight α_{0I} and the angle α_{0T}, are relatively small.

The predictions of the elements of the trajectory are incredibly accurate. It is strange that equations (2.8.6) and (2.8.7) do not give correct estimations, when projectile flies in presence of air drag.

Example 8.1.
Find the super elevation angle for Lapua GB528 19.44 bullet, if the inclined range is 1,200 yards and the inclined angle is E =35 degree. In table 2.1, section 2.2 (first column), we find that the departure angle for horizontal range 1200 yards is the $\alpha_{0T} = 0.6831°$.

Solution
Substituting in, (2.8.14) we find

$$sin(\alpha_{0I}) = sin(\alpha_{0T})\cos(E) = \sin(0.6831)\cos(35°) = 0.0098°.$$

Hence, $\qquad \alpha_{0I} = \arcsin(0.0098) = 0.5596°.$

This is the same result we have got in table 2.1, sec. 2.2 solving differential equation using PC program **RLAPUA16.BAS**, Appendix D. Employing (2.8.5), we get the same result:

$$\alpha_{0I} \approx \alpha_{0T} \cdot \cos(E) = 0.6831° \cdot \cos(35) = 0.5596°.$$

3

Standard Atmosphere in Exterior Ballistics

Introduction

The resistance of air to the flight of a projectile depends on the characteristics of atmosphere, on the physical and aerodynamic characteristics of projectile, as well as on the instantaneous velocity and Mach number.

Characteristics of Atmosphere
The characteristics of atmosphere are the density ρ, the temperature T, the pressure p, the humidity of air, wind.

The exterior ballistics of point-mass projectile considers an ideal atmosphere, where wind is not present, while air density, the pressure and the air temperature decrease with increasing altitude over the sea level, according to the ideal gas law and the hydrostatic equation (section 3.3).

At the sea level, or close to it, for mid-altitude geographic locations, the characteristics of atmosphere are considered constant and unchanged with time and location. Such atmosphere is called Standard Atmosphere.

The standard atmosphere is a hypothetical atmosphere where density, temperature, pressure, humidity of air and speed of sound at the sea level (or close to it), as well as their vertical variation with altitude over the sea level, are established by international agreements.

3.1 Standard Atmosphere

In the practice of exterior ballistics there are in use at least three standard atmospheres: the ICAO atmosphere, the ASM atmosphere, and TSA atmosphere.

ICAO Standard Atmosphere

The International Civil Aviation Organization atmosphere, (ICAO atmosphere) has the following characteristics:

- At the sea level, $y = 0$, the air density is $\rho_{0N} = 1.2251 \ kg/m^3$, the temperature of air is $t_{0N} = 15°\,Celsius$ ($T_{0N} = 288.15 \ Kelvin$), the atmospheric pressure is $p_{0N} = 760mmHg$, the relative humidity of air vapors is $r_H = 0\%$, the speed of sound is $a_{0N} = 340.30m/s$.
- The corresponding virtual temperature is $\tau_{0N} = 288.15°\,K$.
- The wind is absent.

ASM Army Standard Atmosphere

The Army Standard Metro atmosphere (ASM) has the following characteristics:

- At the firing site ($y = 0$), the air density is $\rho_{0N} = 1.2034 kg/m^3$, the temperature of air is $t_{0N} = 15\ °Celsius$, the atmospheric pressure is $p_{0N} = 750\ mm\ Hg$, the relative humidity is $r_H = 78\%$, the speed of sound is $a_{0N} = 341.458\ m/s$.
- The corresponding virtual temperature is approximately $\tau_{0N} = 289.60° K$ [15]
- The wind is absent.

TSA Standard Atmosphere
The Traditional Standard Atmosphere (TSA), mostly used in ex-eastern countries, assumes that the projectile motion is in a standard atmosphere, with the following characteristics:

- At the firing site, , the air density is $\rho_{0N} = 1.205\ kg/m^3$, the temperature of air is $t_{0N} = 15°Celsius$, the virtual temperature is $\tau_{0N} = 289.08K$, and corresponds to the relative humidity $r_H = 50\%$, the atmospheric pressure is $p_{0N} = 750\ mm\ Hg.$, the speed of sound is $a_{0N} = 340.84\ m/s$.
- The wind is absent.

The TSA atmosphere assumes that the projectile is fired at an altitude of 110 meters, over the sea level, where the atmospheric pressure is considered 750 mm Hg.

Note
Bullet companies in USA use ICAO atmosphere, or ASM atmosphere.

Thus, Sierra, Hornady and Barnes use the ASM atmosphere. Lapua and Berger use ICAO atmosphere.
The Russian and other ex-East European countries use the TSA atmosphere.

3.2 Characteristics of Atmospheric Air

In a standard atmosphere, the air is considered an ideal gas at rest that follows the ideal gas law and the hydrostatic equation that describe the decrease of pressure with altitude, i.e. the equations

$$\rho = (1 - 0.3785 \frac{e}{p}) \frac{\mu_A}{R} \frac{P}{T}, \qquad (3.2.1)$$

$$p = (R/\mu_A) \cdot \rho \cdot \tau \qquad (3.2.2)$$

$$dp/dy = -\rho \cdot g, \qquad (3.2.3)$$

where
$R = 8.31441 J \cdot mol^{-1} K^{-1}$, $\mu_A = 28.9644 \cdot 10^{-3} kg/mol^{-1}$ are respectively the universal gas constant and the molar mass of air, while T is the temperature of air in degree Kelvin, τ is the virtual temperature of air (degree Kelvin), ρ is the density of air (kg/m³), and p is the atmospheric pressure in Pa.

(Pressure units: $1 Pa = 1 N/m^2 = 7.5006 \cdot 10^{-3} mm\ Hg$, $1\ mm\ Hg = 133.322\ Pa$.)

Virtual Temperature, Relative Humidity and Vapor Pressure
The virtual temperature τ can be estimated using equation

$$\tau = \frac{T}{1 - 0.3785 \cdot (e/p)}, \qquad (3.2.4)$$

where
    T is the temperature of dry air in degree Kelvin,

$$T = 273.15 + t , \qquad\qquad (3.2.5)$$

e is the partial pressure of saturated water vapor, when relative humidity has a certain percentage value, r_H .

The parameter t is the temperature of dry air in degree Celsius. The partial pressure e and the atmospheric pressure p must be expressed in the same units, i.e. both in Pascal (Pa), or both in mm Hg, or inch Hg.

Employing (3.2.4), the density of humid air (3.2.1) can be expressed through the virtual temperature by the equation

$$\rho = \frac{\mu_A}{R}\frac{P}{\tau} . \qquad\qquad (3.2.6)$$

Partial Vapor Pressure

The saturated pressure of water vapor that corresponds to relative humidity $r_H = 100\%$, can be calculated using the following approximate formulas[16]:

$$e_{100\%} = 7.50187 \cdot e^{19.04 \cdot (1. - 280.07/T)} , \qquad 273.16 \le T \le 327.15 , \quad (3.2.7)$$

or

$$e_{100\%} = 7.50187 \cdot e^{22.024 \cdot (1. - 279.24/T)} , \qquad 255.15 \le T \le 273.15 , \quad (3.2.8)$$

where
    T is the temperature of dry air ($r_H = 0.00\%$) in degree Kelvin.

Multiplying (3.2.7), or (3.2.8) by the relative humidity, r_H , we obtain the corresponding vapor pressure for a given temperature and a given relative humidity.

In standard atmosphere at the sea level, or at shooting ground. The virtual temperature τ_{0N}, can be estimated using (3.2.4):

$$\tau_{0N} = \frac{T_{0N}}{1 - 0.3785(e_{0N} / p_{0N})}, \qquad (3.2.9)$$

where

e_{0N} is the partial pressure of water vapor at the standard temperature $t_{0N} = T_{0N} - 273.15$ and relative humidity r_H.

The saturated pressure of water vapor e can be estimated as well using the data presented in the following table 2.1 (temperature t in degree Celsius, pressure of vapor e in mm. Hg).

Table 2.1 – Saturated Pressure of Water Vapor and (relative humidity, $r_H = 100\%$)

Temp. $°C$	-40	-18	-10	0.0	5.0	10	15
Pressure e	0.15	1.14	1.95	4.58	6.54	9.20	12.70
Temp. $°C$	20	25	30	38	40	54	-
Pressure e	17.54	23.76	31.7	49.2	55.1	115.1	-

If the relative humidity r_H is less than 100%, for example it is $r_H = 50\%$, then, the partial pressure e of water vapor, that corresponds to relative humidity $r_H = 50\%$, is obtained by multiplying the respective value in table 2.1 with $r_H = 50\% = 0.50$

Thus, at the temperature of air $t_0 = 15°Celsius$ (288.15 Kelvin), and $r_H = 50\%$ humidity, the partial pressure of water vapor is $e = 6.35 mmHg$. Substituting the above value in (3.2.4), we find that the virtual temperature is 289.08 Kelvin.

The partial pressure e of the saturated water vapor, that corresponds to an intermediate temperature, which is not displayed in table 2.1 can be found by interpolation.

Speed of Sound

Speed of sound depends as well on the temperature and humidity of air.

The speed of sound a, for a given virtual temperature τ, can be estimated using the equation [17]:

$$a = 20.0469\sqrt{\tau}. \qquad (3.2.10)$$

Example 2.1

The relative humidity of air is $r_H = 78\% = 0.78$, while the temperature and atmospheric pressure are respectively $t_0 = 15°Celsius$ and $p_0 = 750$ mm Hg.

Find the virtual temperature, density of air, and the speed of sound.

(Note that 1 $Pa = 7.5006 \cdot 10^{-3} mm$ Hg; 1 mm $Hg = 133.322$ Pa.)

Solution

Employing equation (3.2.7), we find that the partial pressure of water vapor is approximately

$$e = r_H \cdot e_{100\%} = (0.78) \cdot 7.50187 \cdot e^{19.04(1-280.07/(273.15+15))} = 9.980 \ mm \ Hg$$

The virtual temperature is

$$\tau = \frac{T}{1-0.3785(e/p)} = \frac{273.15+15}{1-0.3785(9.980/750)} = 289.61° \ Kelvin$$

Using the data of table 2.1, we find the same partial pressure and virtual temperature.

Indeed, table 2.1 shows that at temperature $t_0 = 15°Celsius$ the pressure of saturated water vapor is 12.7 mm Hg.

Thus, the pressure of vapor that corresponds to relative humidity 78% is

$$e = r_H \cdot e_{100\%} = (0.78) \cdot (12.7) = 9.906 \ mm \ Hg.$$

The virtual temperature is

$$\tau = \frac{T}{1 - 0.3785(e/p)} = \frac{273.15 + 15}{1 - 0.3785(9.906/750)} = 289.61° \, Kelvin$$

The density of air is

$$\rho = \frac{\mu_A}{R}\frac{P}{\tau} = \frac{28.9644 \cdot 10^{-3}}{8.31441} \cdot \frac{750 \cdot (133.322)}{289.61} = 1.203 \; kg/m^3 \, .$$

The speed of sound is

$$a = 20.0469\sqrt{\tau} = 20.0469 \cdot \sqrt{289.61} = 341.16 \; m/s \, .$$

Example 2.2
Find the virtual temperature, the density of air, and the speed of sound at a firing site where the pressure, the temperature of dry air, and the relative humidity are respectively 760 mm Hg, 25 degree Celsius, and 65%.

Solution
Employing Lewis' Formula (3.2.7) we find that the pressure of water vapors is

$$e = (0.65) \cdot (7.50187 \cdot e^{19.04(1-2800.7/(273.15+25))}) = 15.471 \; mm \, Hg \, .$$

The virtual temperature is

$$\tau = \frac{T}{1 - 0.3785(e/p)} = \frac{273.15 + 25}{1 - 0.3785(15.471/760)} = 300.47 \; °Kelvin \, .$$

The density of humid air is

$$\rho = \frac{\mu_A}{R}\frac{P}{\tau} = \frac{28.9644 \cdot 10^{-3}}{8.31441} \cdot \frac{(760) \cdot 133.322}{300.47} = 1.1748 \; kg/m^3 \, .$$

The speed of sound is

$$a = 20.0469\sqrt{\tau} = 20.0469 \cdot \sqrt{300.47} = 347.491 \; m/s \, .$$

Example 2.3
(a) Find the speed of sound that corresponds to the ICAO atmosphere at the sea level.
(b) Find as well the speed of sound at the ground for the ASM atmosphere.
Consider the shooting ground at the altitude 110 meters above the sea level.

Solution
For the ICAO atmosphere, at the sea level, relative humidity is zero, $r_H = 0$. Using (3.2.8), it results that the corresponding pressure of vapor is zero, $e = 0$. Substituting $e = 0$ in (3.2.9), we find that the virtual temperature is

$$\tau_{0N} = T_{0N} = 288.15 \; {}^\circ K \, .$$

The speed of sound is

$$a = 20.0469\sqrt{\tau} = 20.0469\sqrt{288.150} = 340.30 m/s.$$

For the ASM standard atmosphere, at the firing ground 110 meters over the sea level, we have: $t_{0N} = 15 \; {}^\circ Celsius$, $p_{0N} = 750$, $r_H = 78\%$.

At the temperature $t_{0N} = 15°$, the pressure of saturated water vapor is 12.7 mm Hg. Thus, for the partial pressure of water vapor we find:

$$e = (0.78) \cdot (12.7) = 9.906 \ mmHg.$$

The virtual temperature is

$$\tau = \frac{T}{1 - 0.3785(e/p)} = \frac{288.15}{1 - 0.3785 \cdot (9.906/750)} = 289.60 \ °Kelvin.$$

The speed of sound is

$$a = 20.0469\sqrt{\tau} = 20.0469\sqrt{289.60} = 341.15 m/s.$$

3.3 Change of Air Characteristics with Altitude

The decrease in the characteristics of standard atmospheric air, i.e. the decrease in the virtual temperature τ, density ρ, and pressure p with altitude y over the sea level (until around 11,000 meters) is described respectively by the following equations:

$$\tau = \tau_0 - 0.006328y \qquad (3.3.1)$$

$$\rho = \rho_0 (\frac{\tau_0 - 0.006328y}{\tau_0})^{4.4}, \qquad (3.3.2)$$

and

$$p = p_0 (\frac{\tau_0 - 0.006328y}{\tau_0})^{5.4}, \qquad (3.3.3)$$

where

τ_0, ρ_0, and p_0 are respectively the virtual temperature, the density, and the pressure of air at the firing site, above or at the sea level.

The number -0.006328 is the gradient of vertical temperature, $d\tau/dy = -0.006328$

As a matter of fact, the International Standard Atmosphere (ISA), and ICAO atmosphere considers a temperature gradient $d\tau/dy = -0.0065$.

Relative Density of Air
The relative density of air is the ratio of the density of air at a given altitude over the standard density of air at the sea level, i.e.

$$h(y) = \rho/\rho_{0N}. \tag{3.3.4}$$

Using the equation of (3.3.2), we find that the relative density of air, at the altitude y is

$$h(y) = (\frac{\tau_{0N} - 0.006328 \cdot y}{\tau_{0N}})^{4.4}. \tag{3.3.5}$$

Empirical Formulae
For the relative density of air, we can use the following empirical formulas:

$$h(y) = e^{-0.0001y}, \tag{3.3.6}$$

$$h(y) = 1 - 0.0001 \cdot y \tag{3.3.7}$$

Example 3.1
The temperature of dry air, the pressure and the humidity, at the firing site, are respectively 20 degree Celsius, 750 mm Hg and 75%.
Find the virtual temperature, the density, and the pressure of air at the altitude 600 meters over the firing site.

Solution
At the firing site, we find:
The pressure of water vapors is

$$e = (0.75) \cdot (7.50187 \cdot e^{19.04(1-280.07/(273.15+20))}) = 13.158 \ mm \ Hg \ .$$

The virtual temperature is

$$\tau = \frac{T}{1 - 0.3785(e/p)} = \frac{273.15 + 20}{1 - 0.3785(13.158/750)} = 295.11 \ ^\circ Kelvin \ .$$

The virtual temperature and the relative density of air, at the altitude 600 meters, over the ground, are respectively:

$$\tau = \tau_0 - 0.006328y = 295.11 - 0.006328 \cdot (600) = 291.31 \ ^\circ K \ .$$

and

$$h(y) = (\frac{\tau_{0N} - 0.006328y}{\tau_{0N}})^{4.4} = (\frac{295.11 - 0.006328y}{295.11})^{4.4} = 0.9446 \ .$$

The pressure at altitude 600 m, is

$$p = p_0(\frac{\tau_0 - 0.006328y}{\tau_0})^{5.4} = 750 \cdot (\frac{295.11 - 0.006328 \cdot (600)}{295.11}) = 699.347 \ mm \ Hg.$$

Example 3.2
At sea level, the pressure is 760 mm Hg, while the temperature is 15 degree Celsius (288.15 degree Kelvin).
At what altitude we expect to have a pressure of 750 mm Hg?

Solution
Substituting in (3.3.3), i.e. in

$$p = p_0 \left(\frac{\tau_0 - 0.006328y}{\tau_0}\right)^{5.4},$$

we can write:

$$760 \cdot \left(\frac{288.15 - 0.006328y}{288.15}\right)^{5.4} = 750.$$

Hence, solving for y, we find $y = 111.55\ m$.

Example 3.3

The temperature and pressure, measured at a meteorological station located 500 meters above the sea level, are respectively 10° C and 710 mm Hg, while the relative humidity is 70%.
Find the atmospheric pressure, the virtual temperature, the temperature, and the relative humidity of air at the sea level.

Solution

The temperature measured at the station location in Kelvin is

$$T = 273.15 + 10° = 283.15K.$$

Employing equation (3.3.7) we find that the pressure of water vapors is

$$e_{70\%} = (0.70) \cdot 7.50187 \cdot e^{19.04 \cdot (1.-280.07/T)} =$$

$$= 7.50187 \cdot e^{19.04 \cdot (1-280.07/283.15)} = 6.46\ mm\ Hg.$$

The corresponding virtual temperature is

$$\tau = \frac{T}{1 - 0.3785 \cdot (e/p)} = \frac{283.15}{1 - 0.3785 \cdot (6.46/710)} = 284.13° K.$$

The density of humid air is

$$\rho = \frac{\mu_A}{R} \frac{P}{\tau} = \frac{28.9644 \cdot 10^{-3}}{8.31441} \cdot \frac{(710) \cdot 133.322}{284.13} = 1.161 \ kg/m^3.$$

The temperature, the density, and the pressure of air at the sea level:
Employing (3.3.2), we find that the virtual temperature at the sea level is

$$\tau_0 = \tau + 0.006328 \cdot y = (284.13 + 0.006328 \cdot (500) = 287.29.^\circ K = 14.14 \ ^\circ C$$

Using (3.3.2) and (3.3.3), for the density of atmospheric air, and the pressure at the sea level, we find respectively:

$$\rho_0 = \rho(\frac{\tau_0}{\tau})^{4.4} = 1.161 \cdot (\frac{287.29}{284.13})^{4.4} = 1.173 \ kg/m^3,$$

and

$$p_0 = p(\frac{\tau_0}{\tau})^{5.4} = 710 \cdot (\frac{287.29}{284.13})^{5.4} = 753.69 \ mm \ Hg.$$

3.4 Barometric Formula

The Isothermal Barometric Formula is an approximate equation that expresses the change in pressure with altitude above the sea level, or above a shooting ground, not necessary located at the sea level.

It is obtained assuming that the gravity is constant and the atmosphere is isothermal, i.e. that the atmospheric temperature of air T does not change with altitude (temperature gradient is zero, $dT/dy=0$).

The Isothermal Barometric Formula can be used when the maximum altitude of the projectile trajectory is relatively small, so that the temperature of air can be considered constant for that layer of air the projectile moves into.

Thus, the barometric formula can be used in long range shootings where the trajectory height is relatively small.

The change of pressure with altitude can be approximately estimated using the isothermal barometric formula[18]:

$$p = p_0 \cdot e^{-\mu_A \cdot g \cdot y / (RT)}, \qquad (3.4.1)$$

where p_0 is the pressure at the firing ground, p is the pressure at an altitude "y" over the ground, μ_A is the molar mass of air ($\mu_A = 28.9644 \cdot 10^{-3}$ kg/mol^{-1}), R is the universal constant of ideal gas $R = 8.31441 \; J \cdot mol^{-1} K^{-1}$, g is gravitational acceleration, $g = 9.80665 m / s^2$ and T (assumed constant) is the temperature of dry air in Kelvin.

Substituting in (3.4.1), $\mu_A = 28.9644 \cdot 10^{-3}$, $R = 8.31441 \; J \cdot mol^{-1} K^{-1}$, $g=9.80665 m/s^2$, we can write:

$$p = p_0 \cdot e^{-0.03416 \cdot y / T}. \qquad (3.4.2)$$

The data for the weather, including the atmospheric pressure, are provided by a meteorological station that in general is not located at the firing ground.

The measured data, including the pressure at the altitude of the station location, are reported to the shooting ground, or the measured pressure is brought to the sea level.

Based on the measured pressure p_s at the station location (altitude y_s), we find the pressure p_b at the altitude y_b where the firearm is located.

Using the barometric formula (3.4.2), we can write respectively

$$p_b = p_0 \cdot e^{-0.03416 \cdot y_b / T} ,$$

and

$$p_s = p_0 \cdot e^{-0.03416 \cdot y_s / T} .$$

Dividing the above formulas, for the pressure at the firing site we can write:

$$p_b = p_s \cdot e^{-0.03416(y_b - y_s)/T} . \qquad (3.4.3)$$

A similar formula expresses the change of air density with altitude. Indeed, since at a constant temperature the density is proportional to pressure, considering formulas (3.4.2) and (3.4.3), we can write:

$$\rho = \rho_0 \cdot e^{-0.03416 \cdot y / T} .$$

and

$$\rho_b = \rho_s \cdot e^{-0.03416(y_b - y_s)/T}$$

Since the coefficient $-0.03416/T$ is relatively small we can approximate

$$p_b \approx p_s (1 - 0.03416 \cdot \frac{y_b - y_s}{T}) \qquad (3.4.4)$$

Once the pressure is calculated using (3.4.14), we can use the ideal gas equation

$$\rho = \frac{\mu \cdot p}{R \cdot T} = 0.003484 \frac{p}{T} . \qquad (3.4.5)$$

to find the corresponding density of dry air.

The quantity $h = T/0.03416$ in the right side of above formula is called scale altitude. Scale altitude when we consider the standard temperature 15 degree Celsius (288.15 Kelvin) is

$$h = T/0.03416 = 288.15/0.03416 = 8435.30m.$$

Practical rule to estimate the atmospheric pressure
In practice, to estimate the pressure of air at a given altitude, when it is known the atmospheric pressure at the shooting ground, or at the altitude of meteorological station, we can use the following practical rule.

Considering that the atmospheric pressure decreases approximately 8.50 mm Hg for any 100 meters increase in the altitude, the pressure at an altitude y can be estimated approximately by the formula:

$$p_y = p_0 - 8.5 \cdot (y - y_0)/100, \qquad (3.4.6)$$

where p_0 is the pressure at the altitude y_0.

Thus, if the pressure p_0 at the sea level is 760 mm Hg then at the altitude 100 meters over the sea level it is 751 mm Hg, while at the altitude 1000 meters the pressure is approximately

$$p_y = p_0 - 8.5 \cdot (y - y_0)/100 = 750 - 8.5 \cdot (1000 - 0)/100 = 665 \; mm \; Hg.$$

Note:
In practice of long range shooting, the best way to estimate pressure, temperature is to measure those parameters at the firing site.

Example 4.1

The temperature and pressure, measured at a meteorological station located 500 meters above the sea level, are respectively 10° C and 710 mm Hg, while the relative humidity is 70%.
Find the atmospheric pressure, the virtual temperature, the temperature, and the density of air at the sea level.

Solution
The temperature measured at the station location in Kelvin is

$$T = 273.15 + 10° = 283.15K.$$

Since the atmosphere is isothermal, the same temperature is assumed at the sea level. At sea level, the pressure is

$$p_b = p_s \cdot e^{-0.03416(y_b - y_s)/T} = 710 \cdot e^{[-0.03416 \cdot (0-500)/(283.15)]} = 754.15 \ mm \ Hg.$$

The pressure is quite equal to the value we calculated in example 3.6 (above). Using (3.4.4) we get approximately the same result:

$$p_b \approx p_s(1 - 0.03416 \cdot \frac{y_b - y_s}{T}) = 710 \cdot (1 - 0.03416 \cdot \frac{0-500}{273.15+10}) = 752.83 \ mm \ Hg$$

The virtual temperature at sea level
The temperature of air at sea level is considered equal to the station temperature since the atmosphere is considered isothermal.
Employing equation (3.3.7) we find that the pressure of water vapors at sea level is

$$e_{70\%} = (0.70) \cdot 7.50187 \cdot e^{19.04 \cdot (1.-280.07/T)} =$$

$$= 7.50187 \cdot e^{19.04 \cdot (1-280.07/283.15)} = 6.46 \ mm \ Hg.$$

The virtual temperature is

$$\tau = \frac{T}{1 - 0.3785 \cdot (e / p)} = \frac{283.15}{1 - 0.3785 \cdot (6.46 / 754.15)} = 284.07° K .$$

The density at sea level the density is

$$\rho = \frac{\mu_A}{R} \frac{p}{\tau} = \frac{28.9644 \cdot 10^{-3}}{8.31441} \cdot \frac{(754.15) \cdot 133.322}{284.07} = 1.233 \ kg / m^3$$

Note
We find approximately the same value of air pressure using the approximate formula (3.4.6):

$$p = 710 - 8.5 \cdot (0 - 500) / 100 = 752.50 \ mm \ Hg .$$

4

Elementary Exterior Ballistics

Introduction

Elementary equations, presented in this chapter, are simple approximate equations to predict with great accuracy the elements of the ballistics trajectory.

We show the remarkable Newton-Snell's law and Tangent Law on Similar Trajectories that are some elementary techniques to predict accurately the elements of bullet trajectory in a non-standard atmosphere when we know the standard range table, and vice-versa.

Particularly, those two laws give the possibility to the shooter to easily calculate the departure angle, as well as the aiming angle to zero in the rifle in non-standard atmosphere.

The elementary equations can be used for the horizontal shooting, as well as for the inclined shooting.

Newton's Law and Tangent Law, applied in Exterior Ballistics, are two original achievements presented in my books EBRM (p. 158 and 177), EBNA (p. 101, 125).

4.1 Newton-Snell's Law in Exterior Ballistics

Two "ballistic projectile trajectories" that have geometric, kinematics and dynamic similitude, at their corresponding points, have similar angles of flights and similar scaling elements (velocity, vertex, time, departure angles, and terminal angles) expressed mathematically by similarity equations:

Similarity Equations

- Corresponding coordinates (x, y),

$$x_2 = J \cdot x_1, \qquad y_2 = J^2 \cdot y_1. \qquad (4.1.1)$$

- Corresponding time,

$$t_2 = J \cdot t_1. \qquad (4.1.2)$$

- Corresponding velocity,

$$v_2 = v_1. \qquad (4.1.3)$$

- Corresponding drop

$$\bar{y}_2 = J^2 \cdot \bar{y}_1. \qquad (4.1.4)$$

- Corresponding angles of flights, α_2, α_1, are related according to Newton-Snell's law

$$\sin \alpha_2 = J \sin \alpha_1, \qquad (4.1.5)$$

where $\cos\alpha_2 = \cos\alpha_1$.

Newton-Snell's law expresses the relation between corresponding angles (α_2, α_1) of projectile trajectories, at two corresponding locations (i.e. corresponding points on two respective similitude trajectories).

Scaling factor is denoted by J. Scaling factor that relates the corresponding elements of both trajectories is

$$J = (\rho_{0N}a_{0N})/(\rho_0 a_0),$$

where ρ_{0N} and a_{0N} are respectively the density and speed of sound at a point (x_1, y_1) on the first trajectory, usually in standard atmosphere, while ρ_0 and a_0 are respectively the density and speed of sound at the corresponding point (x_2, y_2) on the second trajectory (non-standard atmosphere).

Scale factor J is the "index of refraction" that represents the ratio of acoustic impedance of air in standard atmosphere and acoustic impedance of air in non-standard atmosphere.

Though the projectile does not actually pass through a physical interface that separates two atmospheres, the equation (4.1.5) is formally the Snell's law of acoustic wave refraction, as it is interpreted by one the greatest scientist Sir Isaac Newton.

As a matter of fact, Snell's Law is related to the refraction of light and has nothing to do with bullets, and their trajectory.
We nominated it "Newton-Snell's Law" because of similarity of formula (4.1.5) and Snell's.

Scaling factor can be written

$$J = (\frac{p_{0N}}{p_0}\sqrt{\frac{\tau_0}{\tau_{0N}}}) = (\frac{p_{0N}}{p_0}\cdot\sqrt{\frac{\tau_{0N}}{\tau_0}}) = (\frac{p_{0N}}{p_0}\frac{a_{0N}}{a_0}). \qquad (4.1.6)$$

As it is shown in EBNA, the set of similarity equations and the Newton-Snell's law equation, applied to motion of projectiles, give remarkably accurate results for the solution of problems of exterior ballistics and to predict the projectile trajectory, not only for small arms but also for the artillery fire.
The accuracy is better for relatively not large launching angles.

In general, the Newton-Snell's law and the other related equations are valid for any two atmospheres, standard or not. **In other words**, Newton-Snell's law relates two non-standard atmosphere trajectories with different acoustic impedances.
If we denote

$$J_{1,2} = (\rho_{01}a_{01})/(\rho_{02}a_{02}) \qquad (4.1.7)$$

the "index of refraction", we can rewrite the Newton-Snell's law for any two similar trajectories, in non-standard atmosphere:

$$\sin\alpha_{02} = J_{1,2}\sin\alpha_{01}. \qquad (4.1.8)$$

The similitude and Newton- Snell's law approach to the solution of exterior ballistics problems is especially very useful in exterior ballistics of small arms to set up the departure angle and the angle of sight, using the individual data obtained in practice of shooting.

Particularly, the use of Newton-Snell's law in exterior ballistics of small arms gives incredibly accurate results.

The following example demonstrates the application of Newton-Snell's law and similitude equations to find the range of a projectile based on the range tables, or shooting data observed during firing tests of a given projectile.

Example 1.1 Mountain Shooting
Use the partial range table 1.1 of 0.338 GB528 Lapua Bullet to find the departure angle needed to hit the target located at a range of $x_{T2} = 1500m$ if the firing site is $y = 1500m$ over the sea level.

- At sea level, the temperature, humidity and pressure are respectively $t = 15°$ Celsius, 0.00% and $p_{0N} = 760$ *mm Hg.*, the corresponding virtual temperature is $\tau_{0N} = 288.15°$. Launching velocity of the projectile is $v_{01} = 830$ *m/s*.

- At altitude $y = 1500$ *m* over the sea level, (non-standard atmosphere) the temperature, pressure and, the virtual temperature (humidity 50%) are respectively:
  $t = 6°$ Celsius, $\quad p_0 = 626$ *mm Hg.*, $\quad e_{050} = 3.536$ *mm Hg.*,
  $\tau_0 = 279.748° K$. Launching velocity is $v_{02} = 830$ *m/s*.

The temperature of propellant charge of cartridge, on both shooting sites is the same, 21.11 degree Celsius.

Table 1.1 - 0.338 GB528 Lapua Bullet

Range [m]	700	800	900	1000	1100
Launching Angle	0.3662	0.4367	0.5126	0.5958	0.6858
Impact Speed	571	539	507	477	448
Time	1.015	1.195	1.386	1.590	1.807
Drop	- 4.494	- 6.139	- 8.086	- 10.439	-13.229

Table 1.1 continue

Range [m]	1200	1300	1400	1500
Launching Angle	0.7838	0.8926	1.013	1.141
Impact Speed	421	395	371	350
Time	2.037	2.283	2.544	2.822
Drop	- 16.52	- 20.368	- 24.858	- 30.074

Solution
Scaling factor is

$$J = \frac{p_{0N}}{p_0}\sqrt{\frac{\tau_0}{\tau_{0N}}} = \frac{760}{626}\sqrt{\frac{279.748}{288.15}} = 1.1962 \ .$$

Using (4.1.1), we find the corresponding range of Lapua projectile at 1500 meters standard atmosphere:

$$x_{T1} = J^{-1}x_{T2} = (1.1962)^{-1} \cdot 1500 = 1253.94 \ m \ .$$

Departure Angle
Using the data of range table of the 0.338 GB528, by interpolation, we find that the departure angle needed to hit the target located at the horizontal range $x_{T1} = 1253.94m$, is $\alpha_{01} = 0.8425°$.

Substituting in Newton - Snell's law, we find that the departure angle (on the mountain site, 1500 meters over the sea level) is

$$\alpha_{02} = \arcsin(J \cdot \sin\alpha_{01}) = \arcsin[1.1962 \cdot \sin(0.8425)] = 1.008° \ .$$

Time of Flight

In the same way, using interpolation, we find that the time of flight to $x_{T1} = 1253.94m$ is $t_1 = 2.168s$.
Thus, time of flight to 1500 meters at the shooting site is

$$t_2 = J \cdot t_1 = 1.19622 \cdot (2.168) = 2.594s \ .$$

Velocity
Using table 1.1, by interpolation, we find that the velocity at $x_{T1} = 1253.94m$ is $v_{T1} = 406.98m/s$. So

$$v_{T2} = v_{T1} = 406.98m/s$$

Projectile Drop

Using interpolation, we find that the drop, at 1253.94 meters, is $\bar{y}_1 = -18.596m$. So, the drop at 1500 meters is

$$\bar{y}_2 = J^2 \cdot \bar{y}_1 = (1.19622)^2 \cdot (-18.596) = -26.61m.$$

The projectile drop can be estimated using departure angle $\alpha_{02} = 1.008°$:

$$\bar{y}_2 = -1500 \cdot \tan(1.008) = -26.392m$$

Note that numerical integration of differential equations shows that elements of Lapua bullet trajectory, at horizontal range $x_{T2} = 1500m$, at altitude $y = 1500$ m are:

Departure angle $\alpha_{02} = 1.007°$, time of flight $t_2 = 2.594s$, velocity $v_{T2} = 406.56m/s$

Note It is obvious that the accuracy of prediction will be better if we would have range tables for ranges that change every 50 meters or less, i.e. for range 50m, 100m, 150m and so on.

4.2 Tangent Law on Similar Trajectories

Newton-Snell's law (4.1.5) is derived applying the theory of dynamic similarity.
There are two essential facts that cannot be explained using only Newton-Snell's law:

1. The Newton-Snell's law is restricted to launching angles α_0 and impact angles α_T , respectively smaller than the critical angles α_{0C} and α_{TC} that correspond to the "total internal reflection" (see example 1.1).

2. The practice of shooting shows that the relation between two corresponding angles of similar trajectories does not depend on the "direction" in which we apply Newton-Snell's law, i.e. from shooting in an atmosphere with greater acoustic impedance to another one with smaller acoustic impedance, or vice versa.

3. Newton-Snell's law (4.1.5) and the equation (4.1.6), respectively

$$\sin\alpha_2 = J\cdot\sin\alpha_1, \text{ and } \cos\alpha_2 = \cos\alpha_1, \quad (4.2.1)$$

are not consistent for relatively large angles.

To make compatible both equations in (4.2.1), we modify the Newton-Snell's law on similar trajectories.
Dividing equations (4.2.1), we can write the "Tangent Law" that relates the corresponding angles of two similar trajectories:

$$\tan\alpha_2 = J\cdot\tan\alpha_1. \quad (4.2.2)$$

All the other similarity equations remain the same, i.e.

$$x_2 = J\cdot x_1, \qquad\qquad y_2 = J^2\cdot y_1. \quad (4.2.3)$$

$$t_2 = J\cdot t_1 \qquad\qquad v_2 = v_1, \quad (4.2.4)$$

The similarity equations, (4.2.3) - (4.2.4), and the "Tangent Law on Similar trajectories" (4.2.2), allow us to determine accurately the elements of trajectory in non-standard-atmosphere using the corresponding elements of the trajectory in standard atmosphere, and vice versa.

Thus, using the range tables that are prepared for a given projectile in standard atmosphere at the sea level (or close to it), we can obtain shooting data or range tables of the same projectile launched in whatever atmosphere.

That is the well-known problem of shooting related with the determination of initial data of shooting in non-standard atmosphere using standard range tables.

Secondly, from the shooting tests performed in non-standard atmosphere, (winter, summer, or high-altitude shooting), we can obtain range tables in standard atmosphere.

That is the well-known ballistics problem of "converting firing data to "sea level standard data".

The Tangent Law and the associated set of similitude equations can be used as well to find the departure angle needed to hit a target located at an inclined range.

It is important to note that the similarity equations and "the tangent law on similar trajectories" can be used to find the elements of trajectory no matter whether:

- The projectile is launched in a non-standard atmosphere at sea level, (for example in winter, or summer time), or
- The projectile is fired at a site located over the sea level (mountain shooting, airborne shooting, etc.).

In general, the "tangent law" is valid for any two atmospheres, standard or not. Thus, for two non-standard atmospheres, the first one with density and speed of sound respectively ρ_{01} and a_{01}, and the second one with density and speed of sound respectively ρ_{02} and a_{02}, we can write:

$$\tan \alpha_{02} = J_{1,2} \tan \alpha_{01}, \qquad (4.2.5)$$

where

$$J_{1,2} = (\rho_{01} a_{01}) / (\rho_{02} a_{02}).\tag{4.2.6}$$

Note that the "tangent law on similar trajectories" (4.2.5), and the related similitude formulae, can be used to construct the range tables of a given projectile in ICAO atmosphere when we know the range tables of the same projectile in ASM atmosphere, or vice versa.

Substituting in (4.2.6):

$\rho_{01} = 1.2034 kg/m^3$, $a_{01} = 341.458 m/s$ (ASM atmosphere),

and
$\rho_{02} = 1.2251 kg/m^3$, $a_{02} = 340.30 m/s$ (ICAO atmosphere),

we find the scaling factor (called as well index of trajectory "refraction"):

$$J_{1,2} = (1.2034 \cdot 341.458) / (1.2251 \cdot 341.458) = 0.9856.$$

Note
The tangent law on similar ballistic trajectories (as well as the Newton-Snell's law) cannot be used to construct the range tables of a projectile in ICAO atmosphere when we know the range tables of the same projectile in TSA atmosphere.

Reason
In TSA atmosphere, the temperature of propellant charge is 15 degree Celsius, while the temperature of propellant charge for shooting in ICAO atmosphere (ASM atmosphere as well) is 21.11 degree Celsius.

The Relevance of Tangent Law on Similar Projectile Trajectories

The similitude equations (4.2.3) – (4.2.4) express the relationships that exist between the elements of two similar trajectories that have different departure angles but respectively identical velocities at corresponding locations on the trajectories, including the initial velocities and terminal velocities.

In long range shooting, since the angles are relatively small, we can use Newton-Snell's law, or Tangent Law.

Thus, for example, using Tangent Law in example 1.1, we find that the departure angle is

$$\alpha_2 = \tan^{-1}(J \cdot \tan\alpha_1) = \tan^{-1}(1.1962 \cdot \tan(0.8425)) = 1.0078° .$$

For relatively small angles, the tangent law can be written:

$$\alpha_2 = J \cdot \alpha_1 . \tag{4.2.7}$$

Thus, for example 1.1, we have

$$\alpha_2 = J \cdot \alpha_1 = 1.1962 \cdot (0.8425°) = 1.0078°$$

Comment
For relatively small angles all similarity equations (4.2.3), (4.2.4), and the Tangent Law (4.2.7) are linearly related.

Compatibility of Similarity Equations and Tangent Law
It is interesting to note the compatibility of the similarity equations.
Indeed:

- Dividing the right equation of (4.2.3) with the left one, we obtain the tangent law (4.2.2).

- Dividing the first equation of (4.2.3) with the first equation of (4.2.4), for the x-components of velocity we can write

$$v_{x2} = v_{x1} \qquad (4.2.9)$$

- Dividing the second equation of (4.2.3) with the first equation of (4.10.4), for the y-components of velocity we can write

$$v_{y2} = J \cdot v_{y1} \qquad (4.2.10)$$

- Using (4.2.9) and (4.2.10), for the projectile velocity we have

$$v_2^2 = v_{x2}^2 + v_{y2}^2 = v_{x1}^2 + J^2 \cdot v_{y1}^2 . \qquad (4.2.11)$$

Thus, we have an equation to estimate the velocity v_2 at a point P_2 on the second similar trajectory, when we know the components of velocity at the corresponding point P_1 on the first trajectory.

The right side of (4.2.11) gives the square of the velocity at the similar point P_1 that corresponds to P_2, i.e.

$$v_1^2 == v_{x1}^2 + J^2 \cdot v_{y1}^2 . \qquad (4.2.12)$$

Note that the similarity equations and the Tangent Law are valid as far as the departure velocity is standard, i.e. the departure velocity remains standard no matter what is the temperature of air at the shooting site.

To keep the departure velocity standard the cartridge should be stored at the temperature of standard atmosphere (21.11 degree Celsius in ICAO or ASM atmosphere, and 15 degree Celsius in ASM atmosphere).

Example 2.1 Critical "Reflection" Angle"

For similitude trajectories of example 1.1, find the critical angle α_c that corresponds to the "total internal reflection".
Scaling factor is $J = 1.1962$.

Solution

To find the critical angle of "total internal reflection" we have to substitute $\alpha_2 = 90°$ in Newton-Snell's law, $\sin\alpha_2 = J\sin\alpha_1$.
Thus, we can write:

$$\sin 90° = 1.1962 \cdot \sin\alpha_1 .$$

Hence, we find that the critical angle is $\alpha_C = 52.718°$.

We cannot apply the Snell's law of ballistic refraction if the departure angle is greater than the critical angle .
Since the impact angle always is bigger than the departure angle, it is obvious that the critical angle is much smaller than the estimated departure critical angle, $\alpha_C = 52.718°$.

Newton-Snell's Law cannot be applied in problems of exterior ballistics for relatively large angles.

So, to avoid errors, for large departure angles, we must always apply the Tangent Law of trajectory refraction and the related similitude formulae.

4.3 Modified Piton-Bressant Trajectory

The simplest parabola that describes the trajectory of projectile flight in presence of resistance of air, but the least accurate, is the 3rd degree parabola that is known also as Piton-Bressant trajectory equation:

$$y = \tan\alpha_0 \cdot x - \frac{gx^2}{2v_0^2 \cos^2 \alpha_0} \cdot (1 + \frac{A_1 \cdot x}{\cos\alpha_0}) , \qquad (4.3.1)$$

Piton-Bressant parabola is shown in Exterior Ballistics with Applications, 3rd. edition, 2011.

To improve the prediction accuracy of the trajectory, in the equation of Piton-Bressant (4.3.1), we introduce a factor "b" that can be determined experimentally together with the constant factor A_1

$$y = \tan\alpha_0 \cdot x - \frac{gx^2}{2v_0^2 \cos^2 \alpha_0} b \cdot (1 + \frac{J^{-1} A_1 \cdot x}{\cos\alpha_0}) , \qquad (4.3.2)$$

where
b and A_1 are parameters that can be estimated using field shooting tests, or reliable standard range tables, possibly Doppler radar data.

We have included as well, in (4.3.2) – (4.3.7) the invers scaling factor J^{-1},

$$J^{-1} = \frac{p_0}{p_{0N}} \sqrt{\frac{\tau_{0N}}{\tau_0}} ,$$

when shooting site is not in a standard atmosphere.
The other equations, associated with modified Piton-Bressant equation, (4.3.2) are:

Angle of flight:

Angle of flight α at any point on the trajectory with abscissa x, or at the impact point x_T is

$$\tan\alpha = \tan\alpha_0 - \frac{gx}{v_0^2 \cos^2 \alpha_0} b \cdot (1 + \frac{3}{2} \frac{J^{-1} A_1}{\cos\alpha_0} x) , \qquad (4.3.3)$$

Projectile Velocity

$$v = \frac{v_0 \cos \alpha_0}{\cos \alpha} b^{-1/2} \cdot (1 + 3 \frac{J^{-1} A_1 \cdot x}{\cos \alpha_0})^{-1/2} . \qquad (4.3.4)$$

Time of Flight

$$t = \frac{b^{1/2}}{v_0 \cos \alpha_0} \int_0^x (1 + 3 J^{-1} A_1 x / \cos \alpha_0)^{1/2} dx . \qquad (4.3.5)$$

Departure angle

$$\tan \alpha_0 = \frac{gx}{2 v_0^2 \cos^2 \alpha_0} b \cdot (1 + \frac{J^{-1} A_1 \cdot x}{\cos \alpha_0}) , \qquad (4.3.6)$$

or

$$\sin(2\alpha_0) = \frac{gx}{v_0^2} b \cdot (1 + \frac{J^{-1} A_1 \cdot x}{\cos \alpha_0}) , \qquad (4.3.7)$$

Note that if the projectile is launched in standard atmosphere then $J^{-1} = 1$.

The trajectory equation (4.3.2) and the related equations are easy to be used in practice of long range shooting.
For the inclined shooting, the departure angle is obtained by substituting in (4.1.2) the coordinates x and y of the impact point, and then solving the resulting equation for the departure angle α_0 (See for analogy example 3.2).

Projectile drop
From trajectory equation (4.3.2), we find the projectile drop, $\bar{y} = y - \tan \alpha_0 \cdot x$

$$\bar{y} = -\frac{gx^2}{2 v_0^2 \cos^2 \alpha_0} b \cdot (1 + \frac{J^{-1} A_1 \cdot x}{\cos \alpha_0}) , \qquad (4.3.8)$$

At the impact point, the horizontal range is $x = x_T$ and the y-coordinate is $y = y_T = 0$.

Shooting with Departure Angle Equal to Zero

The Piton-Bressant equation and all related formulae can be simplified if the departure angle is zero.

Substituting $\alpha_0 = 0°$ in the above equations we have:

The Equation of parabola

$$y = -\frac{gx^2}{2v_0^2} \cdot b \cdot (1 + J^{-1} A_1 x) , \qquad (4.3.9)$$

Note that y is the projectile drop, i.e. $y = \bar{y}$.

Angle of Flight

$$\tan \alpha = -\frac{gx}{v_0^2} b \cdot (1 + \frac{3}{2} J^{-1} A_1 x) , \qquad (4.3.10)$$

Projectile Velocity

$$v = \frac{v_0}{\cos \alpha} b^{-1/2} \cdot (1 + 3 \cdot J^{-1} A_1 x)^{-1/2} . \qquad (4.3.11)$$

Time of Flight

$$t = \frac{b^{1/2}}{v_0 \cos \alpha_0} \int_0^x (1 + 3 J^{-1} A_1 x / \cos \alpha_0)^{1/2} dx . \qquad (4.3.12)$$

Using (4.3.2), or (4.3.9) we can find the parameters b and A_1 .

Example 3.1

For the Lapua GB528 Scenar 19.4 g, Caliber 8.6 mm bullet find b and A_1 considering that shooting is done in Standard ICAO Atmosphere ($J^{-1} = 1$).

Departure angle: $\alpha_0 = 0$. Use the following data:

Range $x_1 = 900m$, drop $\bar{y}_1 = -8.14m$. Range $x_2 = 1200m$, drop $\bar{y}_2 = -16.571m$.

Solution

Substituting in (4.1.9) we have:

$$-8.14 = -\frac{g(900)^2}{2(830)^2} \cdot b \cdot (1 + (1)A_1(900)), \qquad (1)$$

and

$$-16.571 = -\frac{g(1200)^2}{2(830)^2} \cdot b \cdot (1 + (1)A_1(1200)), \qquad (2)$$

Dividing we find that

$$A_1 = 8.56596368 \times 10^{-4} \qquad (3)$$

Substituting (3) in (2), we find

$$b = 0.79726343 \qquad (4)$$

Thus, the trajectory equation (4.3.2), for Lapua bullet, can be written:

$$y = \tan\alpha_0 \cdot x - \frac{gx^2}{2v_0^2 \cos^2\alpha_0}(0.79726) \cdot (1 + \frac{J^{-1}(8.56596 \times 10^{-4} \cdot x)}{\cos\alpha_0}), \qquad (5)$$

While the equation (4.3.9) can be written

$$y = -\frac{gx^2}{2v_0^2} \cdot (0.79726) \cdot (1 + J^{-1}(8.56596 \times 10^{-4}) \cdot x), \qquad (6)$$

In the same way we can write all the other equations.

Example 3.2

For the Lapua GB528 Scenar 19.44 g, Caliber 8.59mm, estimate:

(a) The drop of the bullet at horizontal range $x_T = 1000$ meters

(b) The angle that zeroes the firearm at 1000 meters.

Consider ICAO atmosphere and the initial velocity of bullet $v_0 = 830 m/s$.

Solution

(a) Using (4.3.9) and the parameters A_1 and b, found in example 3.1, we find the projectile drop

$$\bar{y} = -\frac{gx^2}{2v_0^2} \cdot b \cdot (1 + J^{-1} A_1 x) =$$

$$= -\frac{9.80665 \cdot (1000)^2}{2 \cdot (830)^2} \cdot (0.79726) \cdot [1 + (8.56596 \times 10^{-4}) \cdot (1000)] = 10.908 m$$

Note that the drop estimated using differential equations, presented in chapter five, is $\bar{y} = -10.439 m$.

(b) **First method**

Using the non-rigidity principle, we find that the departure angle needed to zero the firearm at 1000 meter is

$$\alpha_0 = \frac{\bar{y}}{x_T} \cdot \frac{180}{\pi} = \frac{10.908}{1000} \cdot \frac{180}{\pi} = 0.625°.$$

Second method

We can find the departure angle using (4.3.6), or (4.3.7). Substituting we have:

$$\sin(2\alpha_0) = \frac{gx}{v_0^2} b \cdot (1 + \frac{J^{-1} A_1 \cdot x}{\cos \alpha_0}) = \frac{9.80556}{830^2} \cdot 0.79726 \cdot (1 + \frac{8.56596 \times 10^{-4} \cdot 1000}{\cos \alpha_0})$$

We can easily solve the above equation using a graphing calculator.

Another simple way to solve the equation (1) is to consider the value of departure angle, on the right side equal to zero, $\alpha_0 = 0$. Thus,

$$\sin(2\alpha_0) == \frac{9.80665 \cdot (1000)}{830^2} \cdot (0.79726) \cdot (1 + \frac{8.56596 \times 10^{-4} \cdot 1000}{\cos(0)}) = 1.20709 .$$

Solving the above equation, we obtain an approximate value:

$$\alpha_0 = \arctan(1.20709)/2 = 0.60354° . \tag{2}$$

Substituting $\alpha_0 = 0.60354$ on the right side of (1) and solving the resulting equation, we get the same value, $\alpha_0 = 0.60354°$.

Note that the value of the departure angle calculated in (2) is close to the value $\alpha_0 = 0.5958°$ obtained solving differential equations (5.1.2).

Substituting in (4.3.8), we find the bullet drop:

$$\bar{y} = -\frac{gx^2}{2v_0^2 \cos^2 \alpha_0} b \cdot (1 + \frac{J^{-1}A_1 \cdot x}{\cos\alpha_0}) =$$

$$= -\frac{9.80665 \cdot (1000)^2}{2 \cdot (830)^2 \cos(0.60354)} \cdot (0.79726) \cdot [1 + \frac{8.56596 \times 10^{-4}(1000)}{\cos(0.60354)}] = -10.536m$$

Note that the estimated drop $\bar{y} = -10.536m$ is quite equal to $\bar{y} = -10.439m$ obtained using numerical integration of differential equations.

The following table 3.1 shows the elements of the Lapua GB528 Scenar bullet trajectory predicted using the equations (4.3.2) - (4.3.8) and the parameters (3) and (4), determined in example 3.1.

Table 3.1 Predicted Data Using Modified Piton-Bressant equations

Range (m)	0	300	600	900	1,200
Velocity (m/s)	830	698.26	583.34	511.22	460.58
Time (s)	0.000	1.404	0.852	1.3937	2.0435
Drop (m)	0.000	-0.644	-3.098	-8.146	-16.571

Note

In all exercises it was assumed that the temperature of the black powder (projectile cartridge) is 21.11 degree Celsius. That means that the initial velocity is equal to standard velocity of Lapua projectile $v_0 = 830 m/s$.

4.4 Exponential Equation of Projectile Trajectory

The trajectory of point-mass projectile can be described relatively accurate employing exponential equation

$$y = \tan \alpha_0 \cdot x - \frac{g}{2v_0^2} (\frac{x}{\cos \alpha_0})^2 e^{kx/\cos \alpha_0}, \qquad (4.4.1)$$

where "k" is a parameter that depends on the projectile and the coordinates of the point of impact, or the horizontal range (see EBRM, p.263, 276).

The parameter k can be estimated using standard range tables, or shooting tests.

In a given standard atmosphere (ICAO, ASM, or TSA), the parameter "k" can be determined using the experimentally obtained coordinates of the point of impact (x_T, y_T) that correspond to a given departure angle α_0 ($\alpha_0 \neq 0$).

Substituting x_T and y_T, respectively for x and y in (4.2.1), we find the parameter "k":

$$k = \frac{\ln[y_T - x_T \cdot \tan\alpha_0) \cdot 2 \cdot v_0^2 \cdot \cos^2\alpha_0) \cdot (9.80665)^{-1} \cdot x_T^{-2}]}{x_T / \cos(\alpha_0)} . \qquad (4.4.2)$$

Note that the projectile drop at range x_T, (point on trajectory with coordinates (x_T, y_T) is

$$\bar{y} = y_T - x_T \tan\alpha_0 . \qquad (4.4.3)$$

The other elements of the exponential trajectory are:

Departure angle that corresponds to zero range $(x_T, \ y_T = 0)$,

$$\tan\alpha_0 = \frac{9.80665}{2v_0^2} \frac{x_T}{\cos^2\alpha_0} e^{k \cdot x_T / \cos\alpha_0} . \qquad (4.4.4)$$

Angle of flight at a given point (x, y) on the trajectory,

$$\tan\alpha = \tan\alpha_0 - \frac{9.80665}{v_0^2 \cdot \cos^2\alpha_0} \cdot x \cdot (1 + \frac{k}{2\cos\alpha_0} x) \cdot e^{kx / \cos\alpha_0} . \qquad (4.4.5)$$

Projectile velocity at the point (x, y) is

$$v = v_0 \frac{\cos\alpha_0}{\cos\alpha} e^{-kx / 2\cos\alpha_0} [1 + 2\frac{kx}{\cos\alpha_0} + \frac{1}{2}(\frac{kx}{\cos\alpha_0})^2]^{-1/2} \qquad (4.4.6)$$

Time of flight, at the point (x, y), is

$$t = \frac{1}{v_0 \cos \alpha_0} \int_0^x e^{kx/2\cos\alpha_0} [1 + 2\frac{kx}{\cos\alpha_0} + \frac{1}{2}(\frac{kx}{\cos\alpha_0})^2]^{1/2} dx \,. \qquad (4.4.7)$$

The definite integral on the right side of (4.4.7) can be estimated using a graphing calculator.

Using (4.4.1), for the projectile **drop** of the projectile,

$$\bar{y} = y - \tan\alpha_0 \cdot x \,,$$

at a point with coordinates (x, y) we can write:

$$\bar{y} = -\frac{g}{2v_0^2}(\frac{x}{\cos\alpha_0})^2 e^{kx/\cos\alpha_0} \,. \qquad (4.4.8)$$

If the bullet is fired horizontally, departure angle $\alpha_0 = 0$, the drop that corresponds to a given range x is

$$\bar{y} = -\frac{g}{2v_0^2} x^2 e^{kx} \,. \qquad (4.4.9)$$

The above equation is obtained substituting $\alpha_0 = 0$ in equation (4.4.8). Using (4.4.9), we find the parameter k,

$$k = \frac{1}{x} \cdot \ln(-\frac{2 \cdot v_0^2 \cdot \bar{y}}{g \cdot x^2}) \,. \qquad (4.4.10)$$

The parameter k depends on the horizontal range x.
If we consider the horizontal range at the impact point, $x = x_T$, then

$$k = \frac{1}{x_T} \cdot \ln(-\frac{2 \cdot v_0^2 \cdot \bar{y}_T}{g \cdot x_T^2}). \qquad (4.4.11)$$

The exponential equation of the trajectory (4.4.1), and the set of corresponding ballistics elements, (equations: (4.4.4), (4.4.5), (4.4.6), (4.4.7), and (4.4.8)) are simple equations and contain only one parameter, "k".
The exponential equation of the trajectory gives acceptable results in predicting the ballistics elements of shooting with small arms.

Comment
Based on non-rigidity and equal drop model, the value of parameter k determined using (4.4.11) is practically equal to the value of k determined using (4.4.2).
Indeed, in (4.4.11) we use the drop that corresponds to range x_T and departure angle $\alpha_0 = 0$, while in (4.4.2) we use the drop

$$\bar{y} = y_T - x_T \cdot \tan(\alpha_0) \qquad (4.4.12)$$

that corresponds to departure angle $\alpha_0 \neq 0°$.
According to the non-rigidity model, the drop is the same no matter what is the departure angle.
If we substitute $\alpha_0 = 0°$ in (4.4.2), we obtain (4.4.11).
The following example (4.1) shows that the value of k in (4.4.10) and (4.4.2) is the same.

Example 4.1
Consider Lapua Scenar GB528 19.44 g. (300 gr.) bullet launched with velocity $v_0 = 830 m/s$, angle $\alpha_0 = 1.147099°$.
(a) Find the parameter "k" if the point of impact that corresponds to the launching angle $\alpha_0 = 1.1471°$ is ($x_T = 1500$, $y_T = 0$).

(b) Find the parameter "k" if the projectile is launched horizontally ($\alpha_0 = 0^0$).

(c) Use value of k obtain in (a) to find all elements of the trajectory.

Solution
Estimating k.
- **(a) Method 1**

    Substituting in (4.4.2): $x_T = 1500$, $y_T = 0$, $\alpha_0 = 1.1471^0$, the initial velocity $v_0 = 830$ and then solving the obtained equation, we find that the value of parameter k is

$$k = 4.1906 \cdot 10^{-4}. \tag{1}$$

- **(b) Method 2**

    Using the principle of non-rigidity, we find that the bullet, drop at 1500 meters is

$$\bar{y} = -x_T \cdot \tan\alpha_0 = -1000 \cdot \tan(1.1471) = -30.035m .$$

Employing (4.4.10) we find that the parameter k is

$$k = \frac{1}{x} \cdot \ln(-\frac{2 \cdot v_0^2 \cdot \bar{y}}{g \cdot x^2}) = \frac{1}{1500} \ln(\frac{-2 \cdot 830^2 \cdot (-30.035)}{9.80665 \cdot (1500)^2}) = 4.1924 \times 10^{-4} .$$

Thus, the equation of the trajectory (4.4.1), for the given Lapua bullet, can be written:

$$y = \tan\alpha_0 \cdot x - \frac{9.80665}{2 \cdot (830)^2}(\frac{x}{\cos\alpha_0})^2 e^{4.1906 \times 10^{-4} \cdot x / \cos\alpha_0} . \tag{2}$$

where $0 \le x \le 1500m$.

(c) Elements of Trajectory

The y-coordinate of bullet at horizontal range $x = 1200m$ is

$$y = \tan(1.1471) \cdot (1200) - \frac{9.80665}{2 \cdot (830)^2} \left(\frac{1200}{\cos(1.1471)}\right)^2 e^{4.1906 \times 10^{-4} \cdot (1200)/\cos(1.1471)} = 7.0794m$$

Substituting the value k (formula 1) and the necessary data in (4.4.8), we find that the drop at $x = 1200m$ is

$$\bar{y} = -\frac{g}{2v_0^2} \left(\frac{x}{\cos \alpha_0}\right)^2 e^{kx/\cos \alpha_0} = -16.9486 \ m$$

Substituting in (4.4.5) we have:

$$\tan \alpha = \tan(1.1471) - \frac{9.80665}{(830)^2 \cdot \cos^2(1.1471)} \cdot (1200) \cdot [1 +$$

$$+ \frac{4.1906 \times 10^{-4}}{2 \cos(1.1471)} (1200)] \cdot e^{4.1906 \times 10^{-4}/\cos(1.1471)} = -0.0153324$$

Hence, we find that the angle of bullet at the point of the trajectory with coordinates (1200, 7.0760m) is

$$\alpha = \arctan(-0.0153324) = -0.878414°.$$

Substituting in (4.4.8), we find that the velocity of the Lapua bullet is

$$v = (830) \frac{\cos(1.1471)}{\cos(-0.878414)} e^{-4.1906 \times 10^{-4}(1200)/2\cos(1.1471)} [1 + 2 \frac{(4.1906 \times 10^{-4}}{\cos(1.1471)} +$$

$$+ \frac{1}{2} (\frac{4.1906 \times 10^{-4}(1200)}{\cos(1.1471)})^2]^{-1/2} = 442.06m/s.$$

Integrating (4.4.9), for example using a graphing calculator, or any mathematics PC software (Maple, Mathematica, etc.), we find that the time of flight is $t = 2.0476$ seconds.

Note In the same way, as in example 4.1, we find all the elements of the trajectory at the points with abscissa 300, 600, ..., 1500.
The obtained results are presented in table 4.1 below.
For comparison, in table 4.2 are given the data obtained by Doppler radar measurements, presented in Wikipedia:

Table 4.1. 0.338 GB528 Scenar bullet. Predicted Data Using Exponential Equation

Range (m)	0	300	600	900	1,200	1,500
Velocity (m/s)	830	695	591	509	442	387
Time (s)	0.000	0.396	0.866	1.414	2.048	2.774
Drop (m)	0.000	-0.727	-3.296	-8.409	-16.949	-30.035

Table 4.2. 0.338 GB528 Lapua Scenar 19.44 g bullet. Doppler radar measurements [19]

Range (m)	0	300	600	900	1,200	1,500
Velocity (m/s)	830	711	604	507	422	349
Time (s)	0.000	0.3918	0.8507	1.3937	2.0435	2.8276
Drop (m)	0.000	-0.715	-3.203	-8.146	-16.571	-30.035

Comment: Comparing table 4.1 and table 4.2, we see that the accuracy of predicted elements of trajectory of the Lapua Scenar GB528 19.44 g (300 gr) bullet, for ranges until 1500 meters are acceptable.
The largest difference in projectile drop (around 0.38 meters) is at the range 1,200 meters.
There are relatively large discrepancies in the estimation of the terminal velocity of the given projectile.

The approximate exponential formula (4.4.1) is easy to be used for any long range shooting since it requires only the launching angle (or drop) that can be obtained experimentally, or from reliable range tables.

Example 4.2
Consider Lapua bullet of example 4.1
(a) Find the launching angle needed to hit the target located at the horizontal range 1200 meters.
(b) Find as well the coordinates of the maximum height of the trajectory.
(c) Find the launching angle that needed to zero the rifle at 1200 meters if the drop of the given projectile at 1,200 meters is 16.571m and not 16.949 m.

Solution
Substituting $y = 0$ and $x = 1200$ in equation (2) of example 4.1, we can write:

$$\tan \alpha_0 = \frac{9.80665}{2 \cdot (830)^2} \frac{(1200)}{\cos^2 \alpha_0} e^{4.18889 \times 10^{-4} \cdot (1200)/\cos \alpha_0}.$$

Hence, we obtain the following equation:

$$\tan \alpha_0 = \frac{0.008541}{\cos^2 \alpha_0} e^{0.50267/\cos \alpha_0}. \tag{1}$$

(a) To find the launching angle we have to solve the transcendental equation (1).

Method 1
The equation (1) can be solved for example using a graphing calculator. Thus, the solution obtained using TI84 Plus is.

$$\alpha_0 = 0.80913°.$$

Method 2

An alternative way is as follows.

Since the launching angle is small, as a first approach, we can consider $\cos\alpha_0 \approx 1$ Substituting $\cos\alpha_0 \approx 1$ in (1), we obtain:

$$\tan\alpha_0 = 0.008541e^{0.50267} = 0.0141196 .$$

Hence,

$$\alpha_0 = \arctan(0.0141196) = 0.80894° . \tag{2}$$

Improving accuracy

Substituting (2) in (1) we have

$$\tan\alpha_0 = \frac{0.008541}{\cos(0.80894)}e^{0.50267/\cos(0.80894)} = 0.014123 . \tag{3}$$

Thus, the departure angle is

$$\alpha_0 = \arctan(0.0141215) = 0.80914° . \tag{4}$$

We can still proceed, in the same way as above, to find another better approximate estimation.

Method 3

Using the principle of non-rigidity of trajectory, considering the calculated drop (example 4.1), we find that the departure angle is

$$\alpha_0 = \tan^{-1}(\frac{\overline{y}_T}{x_T}) = \tan^{-1}(\frac{16.949}{1200}) = 0.80920° . \tag{5}$$

Note that the third method is the simplest one since it requires to calculate the projectile drop using the equation (4.4.8) (see example 6.1), and then, use the principle of non-rigidity of the trajectory.

Using the above estimated angle $\alpha_0 = 0.80920°$ and formula (4.4.1), we can easily verify that the projectile launched with the angle $\alpha_0 = 0.80920°$ will hit the target not at the center, but 0.38 meters above it, i.e. at the point with coordinates (1200m, 0.38m).

Indeed, using (4.4.1), we find:

$$y = \tan\alpha_0 \cdot x - \frac{9.80665}{2\cdot(830)^2}(\frac{1200}{\cos 0.80920})^2 e^{4.1889\times10^{-4}\cdot(1200)/\cos(0.80920)} = 0.38m.$$

(b) Maximum Height
At the maximum height of the trajectory, (x_m, y_m), the angle of projectile flight is $\alpha = 0$. Substituting in (4.4.5) we have:

$$\tan\alpha_0 - \frac{9.80665}{v_0^2\cdot\cos^2\alpha_0}\cdot x_m \cdot (1+\frac{k}{2\cos\alpha_0}x_m)\cdot e^{kx_m/\cos\alpha_0} = 0. \quad (6)$$

The abscissa x_m of the trajectory vertex is obtained solving equation (6), after substituting $k = 4.18889\times10^{-4}$, $\alpha_0 = 0.80914°$, $v_0 = 830$. We can write:

$$0.014123 - 1.423807\cdot10^{-5}\cdot x_m \cdot (1+2.094654\cdot10^{-4}x_m)\cdot e^{4.189308\cdot10^{-4}x_m} = 0$$

The above equation can be solved using a graphing calculator, or any mathematics software, or using the trial and error procedure. The solution of equation is $x_m = 842.74$.

Substituting the above value in (4.4.1), we find that the maximum height of the trajectory is $y_m = 7.195m$.
The coordinates of the trajectory vertex are $(x_m = 842.74m, y_m = 7.195m)$.

(c) Since the real value of the drop at 1200 meters is 16.571m and not 16.949m, we find that the corresponding launching angle is

$$\alpha_0 = \tan^{-1}(\frac{\bar{y}_T}{x_T}) = \tan^{-1}(\frac{16.571}{1200}) = 0.79116°.$$

Example 4.3 For Lapua bullet of example 4.1, find the bullet drop at the following ranges: 1500, 1200, and 900 meters.

Solution
Substituting in (4.4.9), x = 1500 we find that the drop of projectile is

$$\bar{y} = -\frac{g}{2v_0^2}x^2 e^{kx} = -\frac{9.80665}{2\cdot(830)^2}\cdot(1500)^2\cdot e^{(4.1924\times10^{-4}\cdot(1500))} = -30.035m.$$

For the ranges 1200m, and 900 meters we find respectively $y = -16.951m$, and $y = -8.408\ m$.

4.5 Exponential Equation in Non-Standard Atmosphere

The exponential equation of the projectile trajectory in non-standard atmosphere is

$$y = \tan\alpha_0 \cdot x - \frac{g}{2v_0^2}(\frac{x}{\cos\alpha_0})^2 e^{k\cdot J^{-1}\cdot x/\cos\alpha_0} \qquad (4.5.1)$$

where

$$J^{-1} = \frac{p_0}{p_{ON}}\sqrt{\frac{\tau_{ON}}{\tau_0}}. \qquad (4.5.2)$$

The value of k can be easily found if we consider departure angle $\alpha_0 = 0$ (bullet launched horizontally):

$$k = \frac{1}{J^{-1}x_T} \cdot \ln(-\frac{2 \cdot v_0^2 \cdot \bar{y}_T}{g \cdot x_T^2}) \qquad (4.5.3)$$

where x_T and y_T are the coordinates of a terminal point on the trajectory.

Departure angle

$$\tan \alpha_0 = \frac{9.80065}{2v_0^2} \frac{x}{\cos^2 \alpha_0} e^{kJ^{-1}x/\cos \alpha_0} . \qquad (4.5.4)$$

Angle of flight

$$\tan \alpha = \tan \alpha_0 - \frac{9.80665}{v_0^2 \cdot \cos^2 \alpha_0} \cdot x \cdot (1 + \frac{k \cdot J^{-1}}{2\cos \alpha_0} x) \cdot e^{kJ^{-1}x/\cos \alpha_0} . \qquad (4.5.5)$$

Velocity

$$v = v_0 \frac{\cos \alpha_0}{\cos \alpha} e^{-kJ^{-1}x/2\cos \alpha_0} [1 + 2\frac{kJ^{-1}x}{\cos \alpha_0} + \frac{1}{2}(\frac{kJ^{-1}x}{\cos \alpha_0})^2]^{-1/2} \qquad (4.5.6)$$

Time

$$t = \frac{1}{v_0 \cos \alpha_0} \int_0^x e^{kJ^{-1}x/2\cos \alpha_0} [1 + 2\frac{kJ^{-1}x}{\cos \alpha_0}x + \frac{1}{2}(\frac{kJ^{-1}x}{\cos \alpha_0})^2]^{1/2} dx . \qquad (4.5.7)$$

Projectile drop when departure angle is α_0 (including $\alpha_0 = 0$)

$$\bar{y} = y - \tan \alpha_0 \cdot x = -\frac{g}{2v_0^2}(\frac{x}{\cos \alpha_0})^2 e^{k \cdot J^{-1} \cdot x/\cos \alpha_0} . \qquad (4.5.8)$$

Note that in standard atmosphere invers scaling factor is $J^{-1} = 1$. The bullet drop, when the departure angle is zero, is

$$\bar{y} = -\frac{g}{2v_0^2} x^2 e^{k \cdot J^{-1} \cdot x} . \qquad (4.5.9)$$

Example 5.1 Non-Standard Atmosphere
To determine the parameter k, which is present in formulas (4.5.1) - (4.5.8), were performed firing tests, shooting horizontally (launching angle zero degree) on a target located at the distance 1000.
The 0.338 Lapua GB528 Scenar 19.44g bullet was fired with velocity 830m/s in a non-windy weather with temperature 25 degree Celsius, pressure 750mm Hg., humidity 70%.
The results of the tests (range, drop) are: (x = 1000m, drop = y= -10.364m).

(a) Find k.
(b) Estimate the drop and velocity of the projectile flying in ICAO atmosphere, $(J^{-1} = 1)$, at ranges 1000m, 900m, 600 and 300m.
(c) Estimate the drop and velocity of the projectile at ranges 1000m, 900m, 600m and 300m when shooting is done in temperature 0.00 degree Celsius, pressure 750 mm Hg, humidity 50%.

Solution
To the humidity 70% and temperature 25 degree Celsius the pressure of water vapors is $e = 16.632mm\ Hg.$ (table 4.1 section 5.4.2). The corresponding virtual temperature is

$$\tau_0 = \frac{T_0}{1 - 0.3785 \cdot e_0 / p_0} = \frac{273.15 + 25}{1 - 0.3785 \cdot 16.632 / 750} = 298.40°K . \qquad (1)$$

The value of J^{-1} is

$$J^{-1} = \frac{p_0}{p_{0N}} \sqrt{\frac{\tau_{0N}}{\tau_0}} = \frac{750}{760} \sqrt{\frac{288.15}{298.40}} = 0.9699 .$$

Substituting x = 1000, y = -10.364, departure angle $\alpha_0 = 0$ degree, and all the other necessary data in (4.5.1) we can write:

$$-\frac{9.80665}{2(830^2)}(\frac{1000}{\cos 0})^2 e^{k(0.9699)\cdot(1000)/\cos 0} = -10.364,$$

or

$$e^{969.932k} = 1.4561057.$$

Hence,

$$k = \frac{\ln(1.4561057)}{969.932} = 3.87414\cdot 10^{-4}.\qquad (2)$$

The value of $k = 3.87414\cdot 10^{-4}$ is valid for horizontal ranges $0 \le x \le 1000m$.

(b) Substituting $J^{-1} = 1$, $k = 3.87315\cdot 10^{-4}$, $\alpha_0 = 0$ and all the necessary data into (4.5.1), (4.5.5), and (4.5.6) we find:

When x = 1000m: Drop $\bar{y} = -10.48m$, velocity $v = 503m/s$.
When x = 900m: Drop $\bar{y} = -8.17m$,   velocity $v = 526m/s$.
When x = 600m: Drop $\bar{y} = -3.23m$,   velocity $v = 605m/s$.
When x = 300:   Drop $\bar{y} = -0.72m$,   velocity $v = 704m/s$.

(c) When the humidity is 50%, the temperature is 0 degree Celsius, the pressure of water vapors is $e = 2.29mm\ Hg.$. So, we have $J^{-1} = 1.013$.

Substituting $J^{-1} = 1.013$, the value $k = 3.87315\cdot 10^{-4}$, the departure angle equal to zero and all necessary data into (4.5.1) or (4.5.5), and in (4.5.6) we find:

For x = 1000m: Drop $\bar{y} = -10.54m$, velocity $v = 500m/s$.
For x = 900m: Drop $\bar{y} = -8.21m$,   velocity $v = 523m/s$.
For x = 600m: Drop $\bar{y} = -3.24m$,   velocity $v = 603m/s$.

For x = 300:   Drop $\bar{y} = -0.72m$,   velocity $v = 702m/s$.

4.6 Modified Exponential Equation of Trajectory

The exponential equation of bullet trajectory, shown in section 4.5, gives approximate results.

Though the differences between GB528 Lapua bullet drops, estimated using exponential equation (4.5.1) and the respective drops, measured by Doppler radar, are relatively small, the discrepancies in bullet velocity (as well as discrepancy in time) are relatively large.

To reduce those discrepancies, in equation (4.5.1) we introduce a correction factor "b".

Departure Angle Different from Zero
Introducing b, the exponential equation of the projectile trajectory in non-standard atmosphere can be written:

$$y = \tan\alpha_0 \cdot x - \frac{g}{2v_0^2}\left(\frac{x}{\cos\alpha_0}\right)^2 \cdot b \cdot e^{J^{-1}kx/\cos\alpha_0}. \qquad (4.6.1)$$

The correction factor "b" allows us to determine the equation of the trajectory of a bullet using two experimental data (or two data from the range table), i.e. the coordinates of two impact points that correspond to two different ranges.

The other equation related to the equation (4.6.1) are:

$$\tan\alpha_0 = \frac{9.80065}{2v_0^2}\frac{b \cdot x}{\cos^2\alpha_0}e^{kJ^{-1}x/\cos\alpha_0}. \qquad (4.6.2)$$

The equation (4.6.2) can be written in following form

$$\sin(2\alpha_0) = \frac{9.80065}{v_0^2} b \cdot x \cdot e^{kJ^{-1}x/\cos\alpha_0} \qquad (4.6.3)$$

Angle of flight

$$\tan\alpha = \tan\alpha_0 - \frac{9.80665}{v_0^2 \cdot \cos^2\alpha_0} \cdot b \cdot x \cdot (1 + \frac{k \cdot J^{-1}}{2\cos\alpha_0} x) \cdot e^{kJ^{-1}x/\cos\alpha_0}. \qquad (4.6.4)$$

Velocity

$$v = v_0 \frac{\cos\alpha_0}{\cos\alpha} \cdot b^{-1/2} e^{-kJ^{-1}x/2\cos\alpha_0} [1 + 2\frac{kJ^{-1}x}{\cos\alpha_0} + \frac{1}{2}(\frac{kJ^{-1}x}{\cos\alpha_0})^2]^{-1/2}. \qquad (4.6.5)$$

Time

$$t = \frac{b^{1/2}}{v_0\cos\alpha_0} \int_0^x e^{kJ^{-1}x/2\cos\alpha_0} [1 + 2\frac{kJ^{-1}x}{\cos\alpha_0} + \frac{1}{2}(\frac{kJ^{-1}x}{\cos\alpha_0})^2]^{1/2} dx. \qquad (4.6.6)$$

Projectile drop

$$\bar{y} = \tan\alpha_0 \cdot x - y = \frac{g}{2v_0^2}(\frac{x}{\cos\alpha_0})^2 b \cdot e^{k\cdot J^{-1}\cdot x/\cos\alpha_0}. \qquad (4.6.7)$$

Departure Angle Equal to Zero
If the projectile is launched horizontally, departure angle $\alpha_0 = 0$, the projectile drop is

$$\bar{y} = -\frac{g}{2v_0^2} \cdot b \cdot x^2 e^{J^{-1}kx}. \qquad (4.6.8)$$

The other equations related with (4.6.8) can be obtained substituting $\alpha_0 = 0$ in (4.6.2) - (4.6.7). Thus, we have:

Modified Exponential Equation of the Trajectory

$$y = -\frac{g}{2v_0^2} b \cdot x^2 \cdot e^{J^{-1}kx/\cos\alpha_0}. \qquad (4.6.9)$$

which, at the same time, is the projectile drop (4.6.8).

Angle of flight

$$\tan\alpha = -\frac{9.80665}{v_0^2}\cdot b\cdot x\cdot(1+\frac{k\cdot J^{-1}}{2}x)\cdot e^{kJ^{-1}x}.\qquad(4.6.10)$$

Velocity

$$v = \frac{v_0}{\cos\alpha}\cdot b^{-1/2}e^{-kJ^{-1}x/2}[1+2k\cdot J^{-1}x+\frac{1}{2}(k\cdot J^{-1}x)^2]^{-1/2}.\qquad(4.6.11)$$

Time $\quad t = \frac{b^{1/2}}{v_0}\int_0^x e^{kJ^{-1}x/2}[1+2kJ^{-1}x+\frac{1}{2}(kJ^{-1}x)^2]^{1/2}\,dx.\qquad(4.6.12)$

Note that in standard atmosphere $J^{-1}=1$.

Example 6.1
ICAO Atmosphere ($J^{-1}=1$), departure velocity standard, $v_0 = 830 m/s$. Determine the parameters "b" and "k" using the data of the GB528 Lapua Scenar bullet shown in table 4.2, section 4.4, i.e.

Departure angle, $\alpha_0 = 0$.
Range x₁ =1200 meters, drop $\bar{y}_1 = -16.71\,m$,
Range x₂ = 1500 meters, drop $\bar{y}_2 = -30.035\,m$. $\qquad(1)$

Solution
Using equation (4.6.8), for two ranges, we can write

$$\frac{\bar{y}_2}{\bar{y}_1} = (\frac{x_2}{x_1})^2 e^{k(x_2-x_1)}.\qquad(2)$$

Substituting in (1) the data given above, and solving for k, we find:

$$k = 4.94740287\times10^{-4}.\qquad(3)$$

Substituting in (4.6.8), x₂ = 1500, $\bar{y}_2 = -30.035$ m, and k given in (3), and then solving for "b", we find.

$$b = 0.89292742.\tag{4}$$

The table 6.1 is obtained using the equations (4.6.2) - (4.6.8), and the parameters (3) and (4).

Table 6.1 GB528 bullet. Predicted Data Using Modified Exponential Equation

Range (m)	0	300	600	900	1,200	1,500
Velocity (m/s)	830	713	592	499	425	365
Time (s)	0.000	0.396	0.914	1.581	2.430	3.497
Drop (m)	0.000	-0.664	-3.079	-8.036	-16.571	-30.035

Note
Comparing the predicted data of table 6.1 with Doppler radar data (table 4.2), we see that the modified exponential equation of the trajectory gives improved approximation data, close to Doppler radar data.

Example 6.2 Inclined Shooting
Use the modified equation of trajectory related with GB528 Lapua bullet, where
$k = 4.94740287 \times 10^{-4}$, $b = 0.89292742$ (example 6.1), to find the departure angle and the super elevation angle, needed to hit a target located at the inclined range D =OT = $1200m$, if the elevation angle is $E = 30°$.

The firearm is on the ground. Atmospheric conditions are: temperature 0.00 degree Celsius, pressure 750mm Hg, humidity 50%.
Consider the temperature of cartridge 21.11 degree Celsius.
The value of J⁻¹, calculated in example 5.1, is very small:

$$J^{-1} = \frac{p_0}{p_{0N}}\sqrt{\frac{\tau_{0N}}{\tau_0}} = \frac{750}{760}\sqrt{\frac{288.15}{298.40}} = 0.9699 \, .$$

Solution

The coordinates of the target are:

$$x_T = D \cdot \cos(E) = 1200 \cdot \cos(30) = 1039.23$$

and

$$y_T = D \cdot \sin(E) = 1200 \cdot \sin(30) = 600m$$

Substituting in equation (4.6.1), we have

$$600 = \tan\alpha_0 \cdot (1039.23) - \frac{g}{2(830)^2}(\frac{1039.23}{\cos\alpha_0})^2 \cdot (0.89292742) \cdot$$

$$\cdot e^{(0.9699) \cdot (4.9470 \times 10^{-4}) \cdot (103923)/\cos\alpha_0}$$

Solving the above equation with a graphing calculator we find the departure angle

$$\alpha_0 = 30.68058°$$

The super elevation angle is

$$\alpha_{0I} = 0.68058°$$

Note: In table 7.3, for the horizontal range OT= 1200 m, we find that the departure angle is $\alpha_{0T} = 0.7920°$.

Using equation (2.3.1), for the inclined angle we find the same value for the super elevation angle

$$\alpha_{0I} = \alpha_{0T} \cdot \cos E = 0.7920 \cdot \cos(30°) = 0.6859°$$

4.7 Effect of Cartridge Temperature on Projectile Trajectory

The elements of projectile trajectory, presented in range tables of a firearm, are calculated for the standard atmospheric conditions and for some standard ballistic characteristics of the projectile, including the standard departure velocity.

In all approximate equations of projectile trajectories in non standard atmosphere, in preceding sections, we have considered that the initial velocity of the projectile is equal to the standard initial velocity v_0.

So, regardless of the temperature of air at shooting site, the initial velocity is equal to the initial velocity in standard atmosphere.

That is true if we are able to store the cartridges in a temperature that is equal to the temperature of dry air T_0 in standard atmosphere ($T_0 = 21.11$ degree Celsius in ICAIO and ASM, $T_0 = 15$ degree Celsius in TSA).

The temperature of cartridge (propellant charge) must be kept standard in experimental shootings.

The meteorological characteristics of the atmosphere at a shooting site and the projectile ballistic characteristics are usually different from the standard ones.

In general, the marksman has no possibility to store the cartridges in standard temperature T_{0N}.

Temperature of propellant charge is equal to the temperature T at the firing site which, in general, is different from the standard temperature T_{0N}.

The initial velocity V_0 of a bullet in non standard atmosphere can be estimated approximately by the formula[20]

$$V_0 = v_0 \cdot [1 + 0.001 \cdot (T - T_{0N})], \qquad (4.7.1)$$

where T_{0N} and v_0 are respectively the temperature of dry air and the standard initial velocity of projectile.

Note that the coefficient (0.001) in (4.7.1) expresses the efficiency of black powder. The efficiency of black powder changes from one type to another.

For some rifle powders, instead of 0.001 we can use 0.0014, which is obtained using the data given in Ballistica22.[21]

For IMR powders (IMR, Improved Military Rifles), using the data provided by Rinker[22], we conclude that the coefficient of efficiency is 0.001766.

The change of initial velocity, as result of deviation of temperature from the standard one, changes the projectile standard trajectory.

The following examples illustrate the prediction of trajectory using equation (4.7.1).

Example 7.1 Non standard initial velocity
Consider Exercise 5.1, question c.
The temperature of air at shooting site is zero degree Celsius.
The temperature of bullet at the shooting site is equal to zero degree Celsius.
As result the launching velocity of the bullet is

$V_0 = v_0(1 + 0.001(t - 21.11)) = 830 \cdot (1 + 0.001 \cdot (0.0 - 21.11)) = 812.49$ m/s.
We have (see example 5.1):

$$J^{-1} = 1.013 , \; k = 3.87315 \cdot 10^{-4}, \; x = 1000.$$

Substituting the above values in (4.5.9), we find that the drop at 1000 meters is

$$\bar{y} = -\frac{g}{2v_0^2} x^2 e^{k \cdot J^{-1} \cdot x} = -\frac{9.80665}{2 \cdot (812.49)^2} (1000)^2 e^{3.18731 \times 10^{-4} \cdot (1.013) \cdot (1000)} = -10.99m$$

Note that the value of projectile drop predicted using PC program RLAPUA16.bas is (-10.987 m).

Example 7.2 Non-Standard Atmosphere
To determine the parameters b and k of the modified exponential trajectory (4.6.1) there were done some firing tests by shooting horizontally (launching angle zero degree) on two targets respectively at 1000m and 700m.

The 0.338 Lapua GB528 Scenar 19.44g. bullet was fired with velocity 830m/s in a non-windy weather with temperature 25 degree Celsius, pressure 758mm Hg., humidity 70%.
The results of shooting tests (range, drop) are:
($x = 1000$ m, drop $\bar{y} = -10.208$ m), ($x = 700$ m, $\bar{y} = -4.419$ m).

(a) Find the parameters b and k presented in modified exponential trajectory.
(b) Use the obtained formula to estimate the drop of the projectile flying in the ICAO atmosphere ($J^{-1} = 1$), at ranges 1000m, 900m and 600m.
(c) Find the departure angle needed to zero the rifle at 1000 meters if the shooting is in ICAO atmosphere.
(d) Use the obtained equation of modified exponential trajectory to estimate the drop of projectile at 1000 meters when shooting is done in temperature 0.00 degree Celsius, pressure 750mm, humidity 50%.
(e)Find as well the drop and velocity of projectile at horizontal range 900 meter.

The ballistics characteristics (mass and diameter of bullet) are standard, temperature of black powder 25 degree Celsius.

Solution
(a) Using (3.4.7), we find that the pressure of water vapors is

$$e_{70\%} = 0.70 \cdot (7.50187 \cdot e^{19.04 \cdot (1.-280.07/298.15)}) = 16.66mm \ Hg \ .$$

The corresponding virtual temperature is

$$\tau_0 = \frac{T_0}{1 - 0.3785 \cdot e_0 / p_0} = \frac{273.15 + 25}{1 - 0.3785 \cdot 16.661 / 758} = 300.65°K \ .$$
$$(1)$$

The value of invers scaling factor J^{-1} is

$$J^{-1} = \frac{p_0}{p_{ON}} \sqrt{\frac{\tau_{ON}}{\tau_0}} = \frac{758}{760} \sqrt{\frac{288.15}{300.65}} = 0.9764 \ . \qquad (2)$$

The initial velocity of bullet is

$$V_0 = 830 \cdot (1 + 0.001 \cdot (25 - 21.11)) = 833.22m/s \ .$$

Substituting in (4.6.8)

$$\alpha_0 = 0 \ , \ v_0 = 833.22 \ , \ x = 1000 \ , \ \bar{y} = -10.208 \ , \ g = 9.80665 \ ,$$
we have
$$-10.208 = -\frac{9.80665}{2 \cdot (833.22)^2} \cdot (1000)^2 \cdot b \cdot e^{(0.9764) \cdot (1000)k} \qquad (3)$$

In the same way, substituting in (4.6.8),

$$\alpha_0 = 0, \ v_0 = 833.22, \ x = 700, \ \bar{y} = -4.419, \ g = 9.80665,$$

we obtain the following equation:

$$-4.419 = -\frac{9.80665}{2 \cdot (833.22)^2} \cdot (700)^2 \cdot b \cdot e^{(0.9764) \cdot (700)k} . \tag{4}$$

Dividing (3) and (4) we have

$$2.3100 = (1.42857)^2 \cdot e^{(0.9764) \cdot (300)k} . \tag{5}$$

Solving (5) we find that

$$k = 4.22974 \times 10^{-4} . \tag{6}$$

Substituting (6) in (3) we have

$$-10.208 = -\frac{9.80665}{2 \cdot (833.22)^2} \cdot (1000)^2 \cdot b \cdot e^{(0.9764) \cdot (1000) \cdot (0.00042297)} .$$

Solving, we find
$$b = 0.9563333 .$$

Thus, the modified exponential equation of trajectory (4.7.5) is

$$y = \tan\alpha_0 \cdot x - \frac{g}{2(v_0)^2} (\frac{x}{\cos\alpha_0})^2 \cdot (0.9563333) \cdot e^{4.22974 \times 10^{-4} \cdot J^{-1} x / \cos\alpha_0} ,$$

It is valid for $x \le 1000m$.

(b) Substituting above: $J^{-1} = 1$, $\alpha_0 = 0$, $x = 1000$, we find that at $x = 1000$ meters the drop in ICAO atmosphere is

$$\bar{y} = -\frac{9.80665}{2(830)^2}(1,000)^2 \cdot (0.9563333) \cdot e^{4.22974 \times 10^{-4} \cdot (1) \cdot (1000)} = -10.391m.$$

In the same way we find:

Range x = 900m, drop is $\bar{y} = -8.068$;
Range x = 600m, drop $\bar{y} = -3.158$.

(c) Using the non-rigidity principle of bullet trajectory, we find that at 1000 meters the departure angle is

$$\alpha_0 = \frac{|\bar{y}|}{x_T} \cdot \frac{180}{\pi} = \frac{10.391}{1000} \cdot \frac{180}{\pi} = 0.5954°.$$

(d) When the humidity is 50%, the temperature is 0 degree Celsius then the pressure of water vapors is $e = 2.29mm\ Hg$.
Bullet velocity is

$$V_0 = 830 \cdot (1 + 0.001 \cdot (0.00 - 21.11)) = 812.48m/s$$

The virtual temperature is

$$\tau_0 = \frac{T_0}{1 - 0.3785 \cdot e_0 / p_0} = \frac{273.15 + 0}{1 - 0.3785 \cdot 2.29 / 750} = 273.47°K,$$

while

$$J^{-1} = \frac{p_0}{p_{ON}}\sqrt{\frac{\tau_{ON}}{\tau_0}} = \frac{750}{760}\sqrt{\frac{288.15}{273.47}} = 1.0130.$$

Substituting in (7), the invers scaling factor, $J^{-1} = 1.0130$ the initial velocity $V_0 = 812.48m/s$ and $\alpha_0 = 0$ we find that the drop at 1,000 meters is

$$\bar{y} = -\frac{g}{2(812.48)^2}(1000)^2 \cdot (0.9563333) \cdot e^{4.22974 \times 10^{-4} \cdot (1.013) \cdot (1000)} = -10.90m . \quad (8)$$

Note that the drop at 1,000 meters obtained using PC program RLAPUA16.bas is $\bar{y} = -10.987m$.

Note, as well, that there is a small difference of about 0.09 meters.

(e) The drop at 900 meters is

$$\bar{y} = -\frac{g}{2(812.48)^2}(900)^2 \cdot (0.9563333) \cdot e^{4.22974 \times 10^{-4} \cdot (1.013) \cdot (900)} = -8.46m$$

Let's find the angle of flight α at 900 meters. Substituting in (4.4.10) we have:

$$\tan \alpha = -\frac{9.80665}{(812.48)^2} \cdot (0.956333) \cdot 900 \times$$

$$\times [1 + \frac{(4.22974) \cdot 10^{-4} \cdot 1.013}{2}(900)] \cdot e^{(4.22974 \times 10^{-4}(1.013) \cdot (900))} = -0.02243$$

Hence,

$$\alpha = \arctan(-0.02243) = -1.2848° .$$

Velocity

Substituting in (4.6.11) we find that

$$v = \frac{812.48}{\cos(-1.2848)} \cdot (0.956333)^{-1/2} e^{-(4.22974 \times 10^{-4}(1.013) \cdot (900)/2)} \times$$

$$\times [1 + 2 \cdot (4.22974 \cdot \times 10^{-4}(1.013) \cdot 900 + \frac{1}{2}(4.22974 \cdot 10^{-4} \cdot 1.013 \cdot 900)^2]^{-1/2} = 504.44m/s$$

$$v = \frac{v_0}{\cos \alpha} \cdot b^{-1/2} e^{-kJ^{-1}x/2}[1 + 2k \cdot J^{-1}x + \frac{1}{2}(k \cdot J^{-1}x)^2]^{-1/2}$$

4.8 Coriolis Effect

Coriolis force
The rotation of Earth produces the Coriolis force (acting on the projectile),

$$\vec{F}_C = 2 \cdot m \cdot (\vec{v} \times \vec{\Omega}),$$ (4.8.1)

Where
is the angular velocity of the Earth (directed along the Earth axis). The Earth rotates Eastward around the axis (directed North), $\vec{v}$ is the projectile velocity, is projectile mass.

The scalar value of angular velocity of Earth rotation is

$$\Omega = 7.292 \times 10^{-5} s^{-1}$$ (4.8.2)

In general, the Coriolis force produces a vertical deviation of projectile from the center of target and, a constant lateral right-side deviation of projectile, when shooting is in Northern hemisphere.

The Coriolis force modifies the trajectory of projectile curving it on the lateral direction.
To the marksman, it appears that the Coriolis force deflects the point-mass projectile to the right of shooting plane, when projectile is fired on the Northern hemisphere.
The projectile is deflected to the left when the projectile is fired on the South hemisphere.
The Coriolis force applied on the projectile changes the range of shooting as well.

The equation (4.8.1) shows that:

- When shooting is parallel to the equator (Coriolis force zero), there is no lateral deflection.

- When projectile is fired along the meridian, the Coriolis lateral deflection is maximum.

Consider a Cartesian coordinate system where the y-axis is directed along the line that connects the center of the Earth with the muzzle, while x-axis and y-axis are at the horizontal plane at the launching point (muzzle).
The horizontal direction of shooting is along x-axis.

The y-axis makes with the equator plane the angle Λ that is the geodetic latitude.

Consider two other axes: directed North, and directed East.
The direction of shooting is determined by the azimuth which is the angle A that ox_A axis forms with x-axis.

Projections of angular velocity of earth rotation $\bar{\Omega}$ along ox, oy and oz axes are respectively

$$\Omega_x = \Omega \cdot \cos \Lambda \cdot \cos A \,,\ \Omega_y = \Omega \cdot \sin \Lambda , \Omega_z = -\Omega \cdot \cos \Lambda \cdot \sin A \,. \quad (4.8.3)$$

Special cases
- When shooting is along the meridian, Azimuth $A = 0$, $\sin A = 0$. As result,
$$\Omega_z = -\Omega \cdot \cos \Lambda \cdot \sin A = 0 \,.$$

It can be showed that, in this case, the Coriolis force does not affect the range of shooting.

- If the shooting is in Equator and along it, then latitude $\Lambda = 0$ and Azimuth $A = 90°$. Substituting in (4.8.3) we have

$$\Omega_x = 0 \,,\quad \Omega_y = 0 \,,\quad \Omega_z = -\Omega \,.$$

To have an idea about the Coriolis Effect in long range shooting we will show the approximate method presented by Shapiro in his wonderful book Exterior Ballistics, Moscow 1946.

According to Shapiro, change in range Δx_T and the cross deflection Z can be estimated approximately using the following formulae:

$$\Delta x_T = (K \cdot \cos \Lambda \cdot \sin A) \cdot (\frac{g \cdot \Omega \cdot t^3}{6}),\qquad (4.8.4)$$

and

$$Z = (\frac{3 \cdot \sin \Lambda}{\sqrt{\tan \alpha_0 \cdot |\tan \alpha_T|}} - \cos \Lambda \cdot \cos A) \cdot \frac{g \cdot \Omega \cdot t^3}{6},\quad (4.8.5)$$

where

$$K = \frac{3 - \tan \alpha_0 \cdot |\tan \alpha_T|}{\tan \alpha_0 \cdot |\tan \alpha_T|},\qquad (4.8.6)$$

t is the time of flight to range x_T, α_0 and α_T are respectively the departure angle and terminal angle.

or the vertical deviation of the projectile from the center of target we have:

$$\Delta y_T = -\Delta x_T \cdot \tan(\alpha_T).\qquad (4.8.7)$$

Note. At a given latitude, in Northern hemisphere, the lateral deviation of bullet to the right, due to Coriolis Effect, is the same no matter what is the azimuth direction of shooting.

Example 8.1 Coriolis Effect in SI units
Find the vertical deviation and the cross deflection of 0.338 GB528 Lapua Scenar 19.44 g. bullet that results from Coriolis effect at the following range:
$x_T = 1500m$ if departure angle is $\alpha_0 = 1.141°$, terminal angle $\alpha_T = -2.0385°$, time of flight $t = 2.824s$.

Angular velocity of Earth rotation is $\Omega = 7.292 \times 10^{-5} s^{-1}$.

Consider:
(a) Shooting is in latitude $\Lambda = 45°$, azimuth $A = 90°$, shooting East.
(b) Shooting is in latitude $\Lambda = 45°$, azimuth $A = 0°$, Shooting North, along the meridian.
(c) Shooting is in latitude $\Lambda = 45°$, azimuth $A = 270°$, shooting West.

Solution
First, we calculate

$$\frac{g \cdot \Omega \cdot t^3}{6} = \frac{9.80665 \cdot (7.292 \times 10^{-5}) \cdot (2.824)^3}{6} = 0.002684m$$

and

$$K = \frac{3 - \tan\alpha_0 \cdot |\tan\alpha_T|}{\tan\alpha_0 \cdot |\tan\alpha_T|} = \frac{3 - \tan(1.141) \cdot |\tan(-2.0385)|}{\tan(1.141) \cdot |\tan(-2.0385)|} = 4230.842 .$$

(a) Substituting in (4.8.4) and (4.8.7), we find:
Change in range

$$\Delta x_T = (K \cdot \cos\Lambda \cdot \sin A) \cdot (\frac{g \cdot \Omega \cdot t^3}{6}) =$$

$$= (4230.84) \cdot \cos(45) \cdot \sin(90) \cdot (0.002684) = 8.030m$$

Change in vertical direction

$$\Delta y_T = -\Delta x_T \cdot \tan(\alpha_T) = -(8.030) \cdot \tan(-2.0385) = 0.286m.$$

The bullet will hit $|\Delta y_T| = |0.286| = 0.286m$ above the center of the target.

Cross Deflection

$$Z = (\frac{3 \cdot \sin \Lambda}{\sqrt{\tan \alpha_0 \cdot |\tan \alpha_T|}} - \cos \Lambda \cdot \cos A) \cdot \frac{g \cdot \Omega \cdot t^3}{6} =$$

$$= [\frac{3 \cdot \sin(45)}{\sqrt{\tan(1.141) \cdot |\tan(-2.0385)|}} - \cos(45) \cdot \cos(90)] \cdot (4230.84) = 0.214m$$

To hit the center of the target we need to adjust sighting by shifting the aiming point 21.4 centimeter below the center and 21.4 centimeter to the left of the center.

(b) In the same way, for shooting along the meridian, (azimuth $A = 0°$), we find:
$$\Delta x_T = 0m, \quad \Delta y_T = 0m.,$$
and

$$Z = (\frac{3 \cdot \sin \Lambda}{\sqrt{\tan \alpha_0 \cdot |\tan \alpha_T|}} - \cos \Lambda \cdot \cos A) \cdot \frac{g \cdot \Omega \cdot t^3}{6} =$$

$$= [\frac{3 \cdot \sin(45)}{\sqrt{\tan(1.141) \cdot |\tan(-2.0385)|}} - \cos(45) \cdot \cos(0)] \cdot (0.002684) = 0.212m$$

(c) In the same way as in (a) we find:

Change in range

$$\Delta x_T = (K \cdot \cos \Lambda \cdot \sin A) \cdot (\frac{g \cdot \Omega \cdot t^3}{6}) =$$

$$= (4230.84) \cdot \cos(45) \cdot \sin(270) \cdot (0.002684) = -8.030m$$

Vertical deviation

$$\Delta y_T = -\Delta x_T \cdot \tan(\alpha_T) = -(-8.030) \cdot \tan(-2.0385) = -0.286m.$$

Cross Deflection

$$Z = (\frac{3 \cdot \sin \Lambda}{\sqrt{\tan \alpha_0 \cdot |\tan \alpha_T|}} - \cos \Lambda \cdot \cos A) \cdot \frac{g \cdot \Omega \cdot t^3}{6} =$$

$$= [\frac{3 \cdot \sin(45)}{\sqrt{\tan(1.141) \cdot |\tan(-2.0385)|}} - \cos(45) \cdot \cos(270)] \cdot (4230.84) = 0.214m$$

Note the lateral deviation of bullet to the right of shooting plane, due to Coriolis Effect is practically the same no matter what is the azimuth direction of shooting.

Example 8.2 Coriolis Effect in Imperial Units
Find the vertical deviation and the cross deflection of 0.338 GB528 Lapua Scenar 19.44 g. bullet that results from Coriolis effect at the following range:

$x_T = 1200$ yard if departure angle is $\alpha_0 = 0.6831°$, terminal angle $\alpha_T = -1.032°$, time of flight $t = 1.801s$.

Angular velocity of Earth rotation is $\Omega = 7.292 \times 10^{-5} s^{-1}$.
Shooting is in latitude $\Lambda = 45°$, azimuth $A = 0°$ (shooting North), $g = 32.174 ft / s^2$.

Solution
First, we calculate:

$$\frac{g \cdot \Omega \cdot t^3}{6} = \frac{32.174 \cdot (7.292 \times 10^{-5}) \cdot (1.801)^3}{6} = 0.002284\,ft\ .$$

and

$$K = \frac{3 - \tan\alpha_0 \cdot |\tan\alpha_T|}{\tan\alpha_0 \cdot |\tan\alpha_T|} = \frac{3 - \tan(0.6381) \cdot |\tan(-1.032)|}{\tan(0.6381) \cdot |\tan(-1.032)|} = 13967.02\ .$$

Substituting in (4.8.4), (4.8.7) and (4.8.5) we find:

Change in range

$$\Delta x_T = (K \cdot \cos\Lambda \cdot \sin A) \cdot (\frac{g \cdot \Omega \cdot t^3}{6}) =$$

$$= (13,967.02) \cdot \cos(45) \cdot \sin(0) \cdot (0.002284) = 0\,ft$$

Change in vertical direction

$$\Delta y_T = -\Delta x_T \cdot \tan(\alpha_T) = -(0) \cdot \tan(-1.032) = 0\ .$$

The bullet will hit at the center of the target.

Cross Deflection

$$Z = (\frac{3 \cdot \sin\Lambda}{\sqrt{\tan\alpha_0 \cdot |\tan\alpha_T|}} - \cos\Lambda \cdot \cos A) \cdot \frac{g \cdot \Omega \cdot t^3}{6} =$$

$$= [\frac{3 \cdot \sin(45)}{\sqrt{\tan(0.6831) \cdot |\tan(-1.032)|}} - \cos(45) \cdot \cos(0)] \cdot (0.002284) = 0.329\,ft = 3.95in$$

Thus, there is no change in range and no vertical shift over the bull's eye.
In the z-direction the deflection is 3.95 inches.

To hit the center of the target we need to adjust aiming by shifting it 3.95 inch to the left of the target center.

Note: For long range shooting, there is no use to correct the trajectory of point-mass bullet due to Coriolis force.
For artillery projectiles, the correction of trajectory can be done for very long distances.

4.9 Range-Wind and Cross-Wind

An important factor that modifies the projectile motion is wind. Wind is a very complicated motion of an enormous mass of turbulent air.
The velocity of wind changes in magnitude and direction, and depends on the location and time the velocity is measured, as well as on the altitude and the characteristics of the shooting site (field, forest, valley, town, hills, mountains, city etc.).

The trajectory of a projectile flight in a windy weather is different from the trajectory of the same projectile in absence of wind.
Wind deflects the trajectory of the projectile in the direction of motion (x-axis) and in the perpendicular direction (z-axis). As result, the projectile will miss the center of target if the launching angle is set up to hit the target in absence of wind.

Wind is characterized by the velocity-vector $\vec{w}$ that can be seen as composed by the "range wind" $\vec{w}_x$, which blows in the direction of fire, or in opposite direction of fire (along x-axis), and the "cross wind" $\vec{w}_z$ that blows perpendicular to the shooting plane.
Thus,

$$\vec{w} = \vec{w}_x + \vec{w}_z.$$ (4.9.1)

The component of wind in vertical direction (y-axis) is not considered.

In general, the velocity of wind (4.9.1) can change from one shooting to another.
To predict the projectile trajectory in presence of wind, the velocity of wind (4.9.1) is considered constant during the entire trajectory and equal to an average value (expected value).

Projectile Motion in Presence of Wind

The range-wind component of a uniform wind that blows with constant velocity $\bar{w}_x$ changes the velocity of the projectile and the angle of motion. Range wind (as well as cross-wind), acts on projectile during entire trajectory.
Quite immediately, after the projectile leaves the muzzle of firearm, the departure velocity and the departure angle change.
The constant range-wind that blows in the direction of shooting, increases the component of the horizontal velocity of projectile, and decreases the departure angle.
The range wind that blows in the opposite direction of shooting, decreases the initial velocity of projectile, and increases the launching angle.

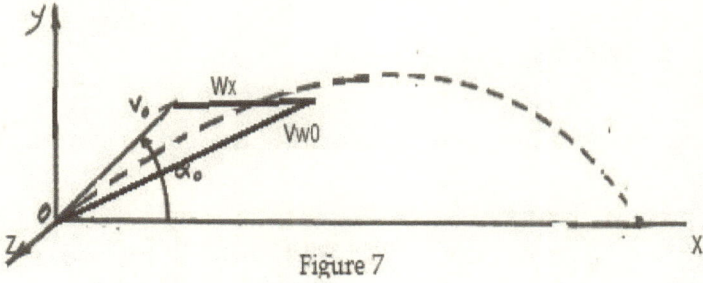

Figure 7

Because of the range-wind, $\vec{w}_x$, the initial velocity of a projectile, quite instantly after it leaves the muzzle of firearm, becomes

$$\vec{v}_{w0} = \vec{v}_0 + \vec{w}_x .\qquad\qquad(4.9.2)$$

The PC programs presented in the book take into account the range-wind and cross-wind deflection.
The deviation of the impact point in the perpendicular direction to the

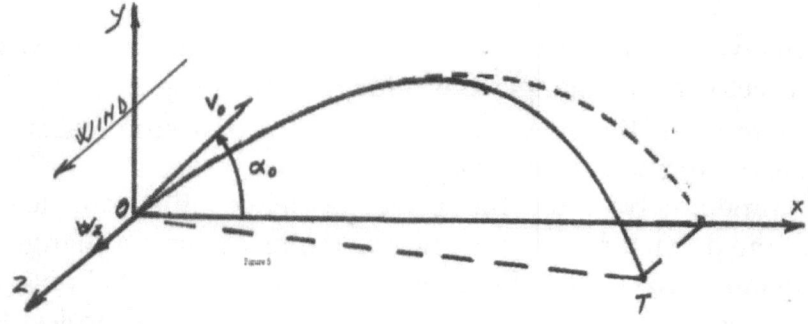

departure plane of projectile, i.e. in the positive direction of the z-axis, or, in opposite direction, is given by Didion's equation

$$z_{Wc} = w_c \left(t_T - \frac{x_T}{v_0 \cos\alpha_0} \right) ,\qquad\qquad(4.9.3)$$

where w_c is velocity of cross wind.

According to McCoy the Didion's classical formula is very accurate.

Note that the PC programs, associated with all my books, estimate the cross-wind deflection based on Didion's formula.

Example 9.1

Consider a 338 Lapua GB528 Scenar 19.44g (300 grain) bullet fired with a departure speed of $v_0 = 830m/s$ at the sea level in the ICAO.

The elements of the trajectory at the horizontal range of 1000 meters are:

Departure angle, $\alpha_0 = 0.5978°$, time of flight, $t = 1.592s$.

Estimate the cross-wind deflection if there is a cross wind component of $w_c = 4m/s$.

Solution

The cross-wind deflection in the positive direction of z-axis is

$$z_{wc} = w_c (t_c - \frac{x_T}{v_0 \cos \alpha_0}) = 4 \cdot (1.592 - \frac{1000}{830 \cdot \cos(0.5978)}) = 1.55m$$

.

The deflection angle at the impact point is

$$\alpha_z = \arctan(1.55/1000) = 0.089°.$$

5

Differential Equations of Exterior Ballistics

"The point-mass trajectory approximation is so generally useful, it may well considered to be the backbone of modern exterior ballistics"

Robert L. McCoy[23]

Introduction

The differential equations of point-mass trajectories can be solved numerically when we know the ballistics coefficient related to reference G_1-function of resistance, or G_7- reference functions, as well as when we know the characteristic G-function of an individual bullet obtained by Doppler radar measurements.

In all three books, EBA, EBNA and EBRM there is shown the way we find the reference standard G-function and characteristic G-function of resistance of a projectile.

The system of differential equations can be solved using numerical methods. Our PC programs use the Runge-Kutta method of integration of differential equations of variable x and relatively big integration steps, 0.1 till around 100 meters.

It is interesting, but surprising to find out that the integration step, we use to solve numerically the differential equations of point-mass projectile trajectory, is relatively big.

"Improved Euler's Method Applied in Exterior Ballistics", presented at the end of the chapter, demonstrates and analyses of that strange finding on big integration steps.

5.1 Differential Equations of Projectile Trajectory

The standard range tables of small arms are constructed using field shooting tests and solving the system of differential equations that describe mathematically the ballistic trajectory of a point-mass projectile (see my books: EBA, EBNA, EBRM, EBLR)[24]

Equations of Variable t , time of flight

$$
\begin{cases}
\dfrac{dv_x}{dt} = -J^{-1} \cdot c \cdot h(y) \cdot G_D(v) \cdot \dfrac{v_x}{v} \\
\dfrac{dp}{dt} = -\dfrac{g}{v_x} \\
\dfrac{dx}{dt} = v_x \\
\dfrac{dy}{dt} = v_y
\end{cases}
\tag{5.1.1}
$$

$(p = \tan\alpha = dy/dx)$

Equations of Variable x

$$\begin{cases} \dfrac{dv_x}{dx} = -c \cdot J^{-1} \cdot h(y) \cdot \dfrac{G_D(v)}{v} \\[2mm] \dfrac{dp}{dx} = -\dfrac{g}{v_x^2} \\[2mm] \dfrac{dt}{dx} = \dfrac{1}{v_x} \\[2mm] \dfrac{dy}{dx} = p \end{cases} \qquad (5.1.2)$$

where $g = 9.80665 m/s^2$ is the constant of gravity,

$$c = \frac{i \cdot d^2}{m} 1000 \qquad (5.1.2a)$$

is the ballistic coefficient (BC), m and d are respectively the projectile mass and the diameter of projectile, i is the form coefficient of the projectile,

$$G(v) = (3.927 \times 10^{-4} \rho_{0N}) v^2 C_D(v/a_{0N}) \qquad (5.1.3)$$

is the function of resistance, $C_D(v/a_{0N})$ is a known reference drag coefficient,

$$h(y) = (\frac{\tau_0 - 0.006328y}{\tau_0})^{4.4} \qquad (5.1.4)$$

is the density function that describes the change of relative air density with the height of flying projectile over the shooting ground.

$$J^{-1} = (\frac{p_0}{p_{0N}} \sqrt{\frac{\tau_{0N}}{\tau_0}}) \qquad (5.1.5)$$

is the invers scaling factor that depends on the virtual temperature and pressure at the shooting site.

The parameters τ_{0N} and p_{0N} are respectively the virtual temperature and pressure of air in standard atmosphere in use, while τ_0 and p_0 are the virtual temperature and the pressure at the shooting site.

The BC, in equation (5.1.2), in imperial units (inch/lb²) is related to BC in SI unites (m/kg²) by the relation:

$$C = 1.4222 / c. \qquad (5.1.6)$$

The quantity,

$$\bar{c} = J^{-1} \cdot c, \qquad (5.1.7)$$

which appears on the right side of (5.1.1), is the **fictive ballistic coefficient**.

The initial velocity of the projectile is equal to the standard launching velocity of the given firearm and projectile, i.e. v_0.

The invers scaling factor can be written:

$$J^{-1} = \frac{p_0 \cdot a_0}{p_{0N} \cdot a_{0N}} \qquad (5.1.8)$$

where a_0 and a_{0N} are the sound speeds in respective atmospheres. Note that we have denoted the invers scaling factor :

$$J^{-1} = 1 / J. \qquad (5.1.9)$$

where $J = p_{0N} a_{0N} / p_0 a_0$.

Note that J^{-1} represents the ratio of **specific acoustic impedances** of air. For the projectile launched at a site with $\rho_0 = \rho_{0N}$ and $a_0 = a_{0N}$ we have $J^{-1} = 1$.

For non-standard atmospheres, $\rho_0 \neq \rho_{0N}$ and $a_0 \neq a_{0N}$, J^{-1} can be less or greater than 1. It becomes zero for the projectile flying in vacuum, $\rho_0 = 0$, $a_0 = 0$.

The specific acoustic impedance is a measure of response (resistance) of the projectile to the applied force.
Thus, J^{-1}, is a characteristic of projectile reaction to the resistance of air in non-standard atmosphere relative to the standard atmosphere.

Standard Ballistic Trajectory
"The prediction of the standard ballistic trajectory is the primary and fundamental problem of external ballistics"[25].

The term **standard ballistic trajectory** indicates the trajectory of the projectile launched in a standard atmosphere, usually in ICAO atmosphere, where temperature of propellant charge is kept standard, 21.11 degree Celsius (70 degree Fahrenheit).

The range table that corresponds to the standard ballistic trajectory is the **standard range table.**

Solving the system of differential equations (5.1.1) we are able not only to predict the standard ballistic trajectory of a projectile but as well the trajectory of point mass projectile in non-standard atmosphere.
Moreover, introducing the wind in system (5.1.1), (5.12) and using our own PC programs, we can predict the projectile trajectory in presence of ballistic wind.

Ballistic Coefficient and Bullet Mass
Any relatively small change in bullet mass (dm/m) changes the
BC of bullet according to the equation

$$c_m = c \cdot (1 - dm/m).$$ (5.1.10)

Thus, a relative increase of $dm/m = 1\%$ in bullet mass, decreases
the BC of GB528 Lapua bullet from $c = 3.796$ into
$$c_m = c \cdot (1 - dm/m) = 3.796 \cdot (1 - .01) = 3.758m^2 kg.$$

The equation (5.1.10) is considered when we predict the projectile
trajectory using PC programs.

5.2 Reference G-Functions of Resistance

Traditionally, in exterior ballistics, to predict the ballistic
trajectory, there are used the well known **Reference Standard G-Functions:** G_1, G_2, G_5, G_6, G_7, G_8, G_{43}, G-Siacci etc.

For long range shooting the accuracy of ballistic trajectory that is
predicted using a reference standard function and the respective
fixed BC is not always accurate, especially for long range
shootings.

**The form coefficient "i" and ballistics coefficient "c" of a bullet,
associated with a reference standard G-function, depend on the
type of the reference standard G-function we adopt to predict
the ballistic trajectory.**

The reference G-functions of resistance that are in use in long range shooting with small arms are mostly G_1 – function and G_7-function.

The G_1 and G_7 functions of air resistance depend on the projectile velocity, density of air ρ_{0N} and speed of sound a_{0N}, which are characteristics of standard atmosphere (ICAO, ASM, TSA).

The analytical reference G_1-function and reference G_7-function, derived using (5.1.3), are respectively[26]:

In ICAO Atmosphere

$$G_1(v) = \begin{cases} 1.0584 \times 10^{-4} v^2 & v \le 256 \ m/s \\ 0.315754v - 78.6769 & 256 < v \le 1000 \ m/s \end{cases} \quad (5.2.1)$$

and

$$G_7(v) = \begin{cases} 5.7679 \times 10^{-5} v^2 & v \le 246 \ m/s \\ 0.152593v - 35.1717 & 256 < v \le 1700 \ m/s \end{cases} \quad (5.2.2)$$

In ASM atmosphere

$$G_1(v) = \begin{cases} 1.00347 \times 10^{-4} \cdot v^2 & for \quad v \le 256 m/s \\ 0.312914v - 79.3976 & for \quad 256 < v \le 1000 \end{cases} \quad (5.2.3)$$

and

$$G_7(v) = \begin{cases} 5.66480 \times 10^{-5} \cdot v^2 & for \quad v \le 256 m/s \\ 0.150355 \cdot v - 34.7319 & for \quad 256 m/s < v \le 1700 \end{cases} \quad (5.2.4)$$

The ballistic coefficient,

$$c = i \cdot \frac{d^2}{m} \cdot 1000, \quad (5.2.5)$$

|208

is related to a given reference G-function and is a characteristic of the projectile (mass m, caliber d, form factor i).

Form factor i and, as result BC of a particular bullet, depend on the reference G-function of resistance, $G_1 = G_1(v)$, or $G_7 = G_7(v)$, as well as on the velocity of projectile.
Thus, if $G(v)$ is the unknown G-function of a particular bullet then the form factor of this bullet with respect to reference G_1-function is

$$i_1 = G(v)/G_1(v),\qquad(5.2.6)$$

or the ratio of respective drag coefficients
$$i_1 = C(v/a_{0N})/C_1(v/a_{0N}).\qquad(5.2.7)$$

The form factor of the given bullet (with unknown drag coefficient) with respect to reference $G_7(v)$ function is

$$i_7 = G(v)/G_7(v),\qquad(5.2.8)$$
or
$$i_7 = C(v/a_{0N})/C_7(v/a_{0N}).\qquad(5.2.9)$$

As the equations (5.2.6) and (5.2.8) show, the form factor of a particular bullet is not a constant but it is a function of the projectile velocity v.

The form factor is nothing much but a function of projectile velocity that matches the unknown G-function of a projectile with a given reference function.

Usually, the form factor i is considered constant and equal to an average value, or a value that is obtained experimentally.

5.3 Characteristic G-functions of Resistance

A G-function of resistance $G(v)$ that is unique for a given type of projectile we will call **Characteristic G-function of resistance**, while the corresponding drag coefficient $C_D(v / a_{0N})$ we will call **characteristic drag coefficient**.

The above nomination allows us to distinguish a characteristic G-function, $G(v)$ of a projectile from a reference G-function (G_1, G_2, ..., G_7). Nowadays, the bullet manufacturers are generating characteristic drag coefficients $C_D(v / a_{0N})$ using Doppler radar measurements for their bullet.
Employing those drag coefficients, we can construct the characteristic G-functions of individual bullets using the equation

$$G(v) = (3.927 \times 10^{-4} \rho_{0N}) v^2 C_D(v / a_{0N}) \qquad (5.3.1)$$

The trend for long range bullets is the use of characteristic drag coefficients $C_D(v / a_{0N})$ obtained using Doppler radar, or other experimental methods.
A series of characteristic drag functions of specific bullets are presented in my book "Exterior Ballistics: The Remarkable Methods", Xlibris, 2014.

In the present book, to predict the elements of the trajectory, we mainly use a characteristic G-functions of air resistance related to an individual bullet.

The form coefficient i, presented in (5.2.5), for a known characteristic G-function of a certain bullet is one.
So, for the corresponding BC we have:

$$c_0 = (1)\frac{d^2}{m}1000.$$

Another important quality of a characteristic G-function (5.1.3) of a specified bullet is that the experimentally generated Doppler radar drag coefficient $C_D(v/a_{0N})$ accounts as well for the jaw-angle of a spinning projectile.

That means that the characteristic G-function **contains as well the contribution of jaw-angle drag coefficient** $C_{D\delta^2}$.

As result, we expect that the **predicted range and some other calculated elements of the bullet trajectory** (in long range shooting) are as accurate as the corresponding elements of the ballistic trajectory calculated using 6-DOF model, or modified point-mass model.

The following table 3.1 is the standard range table of 0.338 Lapua GB528 Scenar 19.44 g. bullet obtained solving the system of differential equations (5.1.2) employing its characteristic G-function:

$$G(v) = 0.141v - 30.031, \qquad (5.3.2)$$

and BC $c_0 = 3.796$.

The standard range table 3.1 is obtained using the PC programs Alapua16.Bas and RLapua16.bas (Appendix D and E, EBLR).

Table 3.1 – Range Table: Bullet 0.338" Lapua GB528 Scenar 19.44g.

Range [m]	100	200	300	400	500
Launching Angle	0.0419	0.0691	0.1353	0.1863	0.242
Impact Speed	791	752	714	677	641
Time	0.122	0.252	0.388	0.532	0.684

| Drop | - 0.074 | - 0.304 | - 0.709 | - 1.307 | - 2.134 |

Table 3.1 - continue

Range [m]	600	700	800	900	1000
Launching Angle	0.3018	0.3662	0.4367	0.5126	0.5958
Impact Speed	606	571	539	507	477
Time	0.844	1.015	1.195	1.386	1.590
Drop	-3.172	- 4.494	- 6.139	- 8.086	- 10.439

Table 3.1 continue

Range [m]	1100	1200	1300	1400	1500
Launching Angle	0.6858	0.7838	0.8926	1.013	1.141
Impact Speed	448	421	395	371	350
Time	1.807	2.037	2.283	2.544	2.822
Drop	- 13.229	- 16.52	- 20.368	- 24.858	- 30.074

Hereafter are listed characteristic G-functions of some bullets.
(1) Characteristic G-function of resistance of Lapua GB528 Scenar 19.44, 8.59 mm, related with ICAO atmosphere is

$$G(v) = \begin{cases} 0.15v - 35.6712 & 280 \leq v \leq 830 \\ 0.02364v - 1.27552 & 136 \leq v < 280 \\ 1.924 \times 10^{-4} v^2 & 0 \leq v < 136 \end{cases} \quad (5.3.2a)$$

For the interval of velocities over around 340 m/s we can use the following characteristic G-function of Lapua GB528 bullet:

$$G(v) = 0.14114 \cdot v - 30.1115, \text{ where } 340 < v < 900 \ m/s. \quad (5.3.3)$$
or

$$G(v) = 0.141v - 30.031, \text{ where } 340 \leq v \leq 830 \ m/s. \quad (5.3.4)$$

The mass and caliber of GB528 Scenar bullet are respectively $m = 0.01944\ kg$, caliber $d = 0.0086\ m$. Departure velocity, $v_0 = 830\ m/s$.

The ballistic coefficient is

$$c_0 = \frac{id^2}{m} 1000 = \frac{(1) \cdot (0.00859)^2}{(0.01944)} 1000 = 3.7957\ m^2/kg. \qquad (5.3.5)$$

(2) The Characteristic G-function of resistance of 0.338 Lapua GB488 Scenar 16.2 gr. (mass $m = 0.0162\ kg$, diameter $d = 0.0086\ m$, ballistic coefficient, $c = 4.5497$, departure velocity, $v = 905\ m/s$) in ICAO atmosphere is

$$G(v) = \begin{cases} 7.9381 \times 10^{-5} v^2 & v \leq 256 \\ 0.146787v - 34.4620 & 256 < v \leq 1021 \end{cases}. \qquad (5.3.6)$$

For the interval of velocities over around 340 m/s we can use the following characteristic G-function of Lapua bullet GB488:

$$G(v) = 0.138776 \cdot v - 28.676. \qquad (5.3.7)$$

where $340\ m/s < v < 1021\ m/s$.
The ballistic coefficient of the bullet is

$$c_0 = \frac{i \cdot d^2}{m} \cdot 1000 = \frac{(1) \cdot (0.00859)^2}{0.0162} \cdot 1000 = 4.555\ m^2/kg$$

(3) **Characteristic G-function, Caliber 0.308, 168 Grain Sierra International bullet**.

The characteristic G-function of Caliber 0.308, 168 Grain Sierra International bullet (mass, $m = 0.010886\ kg$; diameter,

$d = 0.0078232\ m$; BC $c_0 = 5.6325$; (i=1); muzzle velocity, $v_0 = 792.48\ m/s$.), ICAO atmosphere is

$$G(v) = \begin{cases} 6.73536 \times 10^{-5} v^2 & v \leq 256 \\ 0.179117v - 46.77305 & v > 256 \end{cases}. \qquad (5.3.8)$$

(4) Characteristic G-function of Resistance of M118LR Bullet

The characteristic G-function of resistance of M118LR bullet (mass 0.01134 kg, diameter $d = 0.0078232\ m$, departure velocity, $v_0 = 884\ m/s$, is

$$G(v) = \begin{cases} 4.81097 \times 10^{-5} v^2 & v \leq 256 \\ 0.178659v - 46.77305 & v > 256 \end{cases} \qquad (5.3.9)$$

Ballistic coefficient is

$$c_0 = \frac{i \cdot d^2}{m} 1000 = \frac{(1) \cdot (0.0078232)^2}{(0.01134)} 1000 = 5.3970\ m^2/kg,$$

(5) Characteristic G-function of Resistance of 0.50 Caliber, MK211 Bullet

The ICAO atmosphere G-function of resistance of the 0.50 - Caliber MK211 Bullet (mass $m = 0.043441\ kg$, caliber $d = 0.01295\ m$, ballistic coefficient, $c_0 = 3.8627$, departure velocity, $v_0 = 827.53\ m/s$) is

$$G(v) = \begin{cases} 5.14774 \times 10^{-5} v^2 & v \leq 256 \\ 0.185906v - 47.78576 & v > 256 \end{cases}. \qquad (5.3.10)$$

(6) Characteristic G-function of Resistance of 300 - Grain .338 - .416 Bullet

The G-function of 300 - Grain .338 - .416 bullet (mass $m = 0.01944$ kg, caliber $d = 0.00859$ m, ballistic coefficient, $c_0 = 3.7914$, departure velocity, $v_0 = 927.40$ m/s), in ICAO atmosphere is

$$G(v) = \begin{cases} 7.07213 \times 10^{-5} v^2 & v \le 256 \\ 0.149642v - 34.00946 & v > 256 \end{cases}. \qquad (5.3.11)$$

5.4 Characteristic G-function Versus Reference G-function

The accuracy of projectile trajectory predicted using reference G_1, or G_7 functions of resistance and a fixed ballistic coefficient is not always satisfactory. To illustrate and explain the discrepancies in projectile trajectory, let's consider the vector differential equation that describe the projectile trajectory in standard atmosphere

$$\frac{d\vec{v}}{dt} = \vec{g} - c \cdot h(y) \cdot G_D(v) \frac{\vec{v}}{v}. \qquad (5.4.1)$$

The acceleration (5.4.1), and as result, the corresponding trajectory, depends on BC and the G-function of resistance.
As the above equation show, there are two main reasons why the trajectories predicted using two different G-functions are not the same:

- The characteristic G-function of a given projectile is not identical to the reference G-function (G_1, or G_7).
- A fixed BC of a bullet, related to a reference G-function, can not adjust the reference standard G-function to match perfectly the characteristic G-function of the given projectile.

Indeed, let's consider **0.338 Lapua GB528 Scenar 19.44 g. bullet,** the reference functions G_1 and G_7, respectively given in formulas (4.2.1) and (4.2.2), as well as the characteristic G-function of resistance given in (4.3.4), i.e. respectively

$$G_1(v) = 0.315754 \cdot v - 78.6769 \qquad (5.4.2)$$

where $256 < v < 1000 m/s$,

$$G_7(v) = 0.152593 \cdot v - 35.1717, \qquad (5.4.3)$$

where $246 < v < 1700 m/s$,

$$G(v) = 0.141 \cdot v - 30.031, \qquad (5.4.4)$$

where $340 < v < 900 \ m/s$.

Instead of (5.4.4), we can apply (5.3.3), i.e.

$$G(v) = 0.14114 \cdot v - 30.1115, \qquad (5.4.5)$$

where $340 < v < 900 \ m/s$.

In table 5.1 there are shown the values of the G-functions (5.4.5), (5.4.4) and (5.4.3) for a range of values of projectile velocity. For example, for velocity $v = 571$ we have:

$$G(v) = 0.14114 \cdot v - 30.1115 = 0.14114 \cdot (571) - 30.116 = 50.48$$

$$G_7(v) = 0.152593v - 35.1717 = 0.152593 \cdot (571) - 35.1717$$
$$= 52$$

Form Coefficient

$$i_7 = G(v)/G_7(v) = G(571)/G_7(571) = 0.972$$

As we can see from table 5.1, the values of the G_7-function are a close match to the values of characteristic G-function of Lapua bullet.

That is the reason why G_7-function is considered close to the drag function of modern long range bullets.

Form Coefficient and BC

In the last two columns of table 5.1 there are shown the form coefficients i_7 and i_1 estimated using formulas (5.2.8) and (5.2.6), i.e.

$$i_7 = G(v)/G_7(v) \quad \text{and} \quad i_1 = G(v)/G_1(v) .$$

The average form coefficient of Lapua GB528 bullet, with respect to G_7 and G_1, is respectively

$$i_7 = 0.9847, \quad i_1 = 0.5133$$

The corresponding BCs are

$$c_7 = i_7 \cdot \frac{d^2}{m} 1000 = i_7 \cdot c_0 = (0.9847) \cdot (3.796) = 3.738 ,$$

and

$$c_1 = i_1 \cdot \frac{d^2}{m} 1000 = i_1 \cdot c_0 = (0.5133) \cdot (3.796) = 1.948 .$$

Note that Wikipedia[27], for GB528 Lapua bullet employs the following BCs: $c_7 = 3.7724$ and $c_1 = 1.8117$.

Lapua Company considers $c_7 = 3.7036 \ m^2/kg$ ($C = 0.384 \ lb/in^2$)[28].

The different values presented above are result of different approximate characteristic G-functions used by different authors (companies).

Table 5.1

Range	Velocity	G-Lapua	G_7	G_1	i_7	i_1
0	830	86.999	91.114	185.468	0.951	0.474
100	791	81.500	85.258	172.864	0.953	0.476
200	752	76.001	79.401	160.260	0.955	0.479
300	714	70.643	73.695	147.978	0.957	0.481
400	677	65.426	68.138	136.020	0.960	0.484
500	641	60.350	62.732	124.385	0.963	0.488
600	606	55.415	57.476	113.074	0.967	0.492
700	571	50.480	52.220	101.762	0.972	0.497
800	539	45.689	47.415	91.420	0.976	0.502
900	507	41.456	42.610	81.078	0.983	0.509
1000	477	37.226	38.104	71.382	0.990	0.517
1100	448	33.137	33.750	62.010	0.998	0.528
1200	421	29.330	29.695	53.284	1.009	0.541
1300	395	25.664	25.791	44.881	1.022	0.557
1400	371	22.280	22.186	37.124	1.039	0.579
1500	350	19.319	19.033	30.337	1.059	0.607
					I_7 avrg. 0.9847	I_1 avrg. 0.5133

In table 5.2 is given the drop and the impact velocity of GB528 Lapua bullet predicted by solving system (5.1.1), employing G_7 or G_1, as well as corresponding BCs, $c_7 = 3.738$ and $c_1 = 1.948$.

The predicted values are obtained using the PC program RPROJ16.BAS which employs the reference G-functions (5.4.4), or (5.4.3), (Appendix E, EBLR).

For comparison, in the last two rows of table 4.2 is given the impact speed and drop presented in Wikipedia External Ballistics[29]

Table 5.2

Range [m]	300	600	900	1200	1500
Impact Speed (G1)	706	592	493	411	348
Drop (G1)	- 0.714	- 3.222	-8.279	-17.017	-31.017
Impact Speed (G7)	710	600	501	416	349
Drop (G7)	-0.711	-3.191	-8.157	-16.690	-30.379
Impact Speed Lapua	711	604	507	422	349
Drop GB528	- 0.715	-3.203	-8.146	- 16.571	-30.035

For long range shooting, table 5.2 shows that the values of velocity and drop, predicted using G7 and the fixed BC, $c_7 = 3.738$, are acceptable.

For long ranges, the velocity and drop predicted using G1 and $c_1 = 1.948$ are not accurate enough.

Table 5.3 Median BC

	G7	G1
Mean	0.9847	0.5132
Median	**0.9740**	**0.4995**
St Deviation	0.033	0.0395
Range	0.108	0.132

In table 5.3 there are shown the statistics of form coefficients related with the data of table 5.1.

As we have shown, the projectile trajectory, predicted using G1 and the fixed BC, $c_1 = 1.948$, is not accurate.

An alternative method to improve the accuracy is to use the median value of the form coefficients instead of the respective mean value, i.e. the form coefficients $i_7 = 0.9740$ and $i_1 = 0.4995$. The corresponding BCs are respectively:

$$c_7 = i_7 \cdot \frac{d^2}{m} 1000 = i_7 \cdot c_0 = (0.9740) \cdot (3.796) = 3.697$$

and

$$c_1 = i_1 \cdot \frac{d^2}{m} 1000 = i_1 \cdot c_0 = (0.4995) \cdot (3.796) = 1.896 \,.$$

In the following table 5.4 are given the GB528 Lapua bullet velocity and drop predicted using reference G_7 -function, reference G_1-function and the respective median values of BC s, $c_7 = 3.697$, $c_1 = 1.896$.

Table 5.4

Range [m]	300	600	900	1200	1500
Impact Speed (G1)	709	598	500	419	356
Drop (G1)	- 0.712	- 3.201	-8.187	-16.74	-30.373
Impact Speed (G7)	712	602	504	419	352
Drop (G7)	-0.710	-3.183	- 8.121	-16.583	-30.118
Impact Speed (Lap)	711	604	507	422	349
Drop GB528	- 0.715	-3.203	-8.146	- 16.571	-30.035

For comparison, in the last two rows of table 5.4, is given the impact speed and the drop according to (end note 2).

The comparison shows that the accuracy in bullet GB528 drop and velocity predicted using the median value of BC is satisfactory and very accurate when we use reference G_7-function.

Notice that not only the bullet drops are equal, but also the corresponding velocities are quite equal.

Interval BC

Lapua Company and Sierra use the **Interval Ballistics Coefficients** to improve the data predicted obtained by using reference G_1-function and a fixed BC[30].

To illustrate the Interval BC, we consider 0.338 GB528 Lapua Scenar bullet and the fixed BC, $c = 1.8117 m^2 / kg$ ($C = 0.785 in^2 / lb.$) that corresponds to G_1 function of resistance [Wikipedia].

improve the prediction accuracy of the trajectory of 0.338" GB528 Lapua Scenar bullet, we can consider the average form coefficient in smaller intervals of velocities: 830 -714, 714 - 606, 606 - 507, 507 - 421, 421 - 350 (see table 4.1).

In table 5.4 is given the ballistic coefficients versus the velocity interval estimated using the data from table 5.1, i.e. the averages of any four form coefficients.
Thus, for example, using data in table 5.1, we find the form coefficient:

$$i_1 = (0.474 + 0.476 + 0.479 + 0.481)/4 = 0.4777$$

that corresponds to the interval of velocities 830 – 714.
The interval form coefficients and interval BCs are presented in table 5.5.
In the last two rows of table 5.5 are given the velocity and the bullet drop predicted using the calculated Interval BCs and G_1-function.

Table 5.5 Interval Ballistic Coefficient of Lapua GB528 Scenar bullet

Velocity	830 - 714	714 - 606	606 - 507	507 - 421	421 - 350
i_1	0.4777	0.4863	0.500	0.5238	0.5710
BC_1	**1.813**	**1.846**	**1.898**	**1.988**	**2.168**
Range	300	600	900	1200	1500
Impact Speed	714	605	506	420	349
Drop	-0.709	-3.169	-8.07	-16.462	-29.935

It can be easily verified (comparing data of table 5.5 with corresponding data of table 5.1) the good accuracy of the trajectory predicted using reference G_1-function and the interval BCs.
Comparing the data of the two last rows of table 5.5 with the corresponding data given in the Wikipedia table, we see that there is a significant improvement due to the interval BC.

Comment
As it is shown above, the variety of fixed BCs (or interval BCs) that can be used to predict with acceptable accuracy the bullet trajectory using G_1, or G_7 reference functions spells out the useless efforts to measure "accurately" a "manufacturer exaggerated BC" [31].

Above all, a BC measured within a short interval of 200 feet, or 300 feet, cannot represent the whole ranges of long range shootings.

Estimating Ballistic Coefficient. Method 2
Consider the equation (1.12.5) of EBRM book, i.e. the average BC,

$$\bar{c}_k = \frac{\bar{v}_{k-1}^2 \cdot \ln(v_{k-1}/v_k)}{h(y) \cdot G_D(\bar{v}_{k-1}) \cdot (x_k - x_{k-1})}, \qquad (5.4.6)$$

in the interval (v_{k-1}, v_k).

We assume that we know the projectile velocity at different ranges, let's say every 100 meters, from zero till a certain range. We denote $\bar{v}_{k-1}$ the average velocity in the interval (v_{k-1}, v_k).

In the last two rows of the table 5.6 are shown respectively the average values of BC and the related standard deviation values.

For example, for the GB528 Lapua bullet ($G(v) = 0.141 \cdot v - 30.031$) substituting in (5.4.6) we find the average value of BC in the interval of velocities 791-830:

$$c_{100} = \frac{((830+791)/2)^2 \cdot \ln(830/791)}{(1) \cdot (0.141 \cdot (830+791)/2 - 30.031) \cdot (100-0)} = 3.7526.$$

Table 4.6

Range	Velocity	G-Lapua	G_7	G_1
		Ballistic Coefficient		
0.00	830	3.753	3.572	1.784
100	791	3.822	3.645	1.825
200	752	3.800	3.633	1.824
300	714	3.783	3.628	1.826
400	677	3.773	3.629	1.834
500	641	3.771	3.640	1.847
600	606	3.891	3.772	1.923
700	571	3.684	3.588	1.840
800	539	3.830	3.751	1.936
900	507	3.753	3.700	1.926
1000	477	3.814	3.790	1.992
1100	448	3.757	3.770	2.005
1200	421	3.859	3.918	2.116
1300	395	3.836	3.951	2.176
1400	371	3.641	3.817	2.154
1500	350			
	Mean value	**3.784**	**3.682**	**1.934**
	Std. Deviation	0.064	0.079	0.129

Using (5.4.6) and the velocities shown in table 5.1, we have estimated the ballistic coefficient related to the GB528 Lapua characteristic G-function (third column of table 5.6), G_7-function (column 4) and G_1- function (column 5).

The values of BC we consider to predict the trajectories of the given Lapua bullet are the averages:

$$c_7 = 3.682, \qquad c_1 = 1.934 .$$

5.5 **Measurement of Muzzle Velocity**

The initial velocity v_0, called as well the muzzle velocity, is an important parameter needed to predict the ballistics trajectory of a projectile.

As we have seen in paragraph 1.1, "the **initial velocity** of a bullet (projectile) is a **fictive velocity** that makes the projectile follow the real trajectory if, at the muzzle, the projectile is launched at the **fictive velocity**, assuming that at the muzzle the powder gases cease to increase exit velocity of bullet beyond the muzzle.

The initial velocity of bullets can be measured easily using the chronographs that are already available in the market.
Two chronographs measure the time Δt_{12} that the bullet passes a short distance Δx_{12} between two screens that are located at least around 4 meters from each other.
The first screen is close to the muzzle in such a distance that the expanding gases cannot put it out of order. The first screen should be at a distance where the expanding gases has ceased to increase the bullet velocity (which can be from few centimeters to some few meters.

For example, Sierra company   describes the measurement of "muzzle" velocity using two chronographs located around 10 feet from each other. (see
http://www.exteriorballistics.com/ebexplained/5th/232.cfm).

At Sierra experiment there is also a shield screen close to muzzle, to protect the first screen from expanding gases.
That shield might reduce the velocity of bullet.

 A similar method of measurement is described by Bryan Litz (see Applied Ballistics for Long Range Shooting, 2009, page 33)
As a matter of fact, the method described by Sierra and Litz measures the average velocity

$$\bar{v} = \frac{\Delta x_{12}}{\Delta t_{12}} , \qquad (5.5.1)$$

at the point in the middle of the distance between two screens.

So, the actual method measures an average velocity at a distance x from the muzzle. Thus, the measured average velocity (5.5.1) is a function of the distance from the muzzle to the midpoint between screens, i.e.

$$\bar{v} = v(x). \qquad (5.5.2)$$

The average velocity depends on the midpoint. It is not constant, not consistent, and so not reliable to represent the muzzle velocity.

Using the measured average velocity (5.5.1) we are able to find the muzzle velocity.
Indeed, let's consider the vector equation (2.1.14) of "Exterior Ballistics: The Remarkable Methods" (Klimi, G., Xlibris, 2014),

$$\frac{d\vec{v}}{dt} = g - c \cdot J^{-1} h(y) \cdot G_D(v) \cdot \frac{\vec{v}}{v}, \qquad (5.5.3)$$

where

$$J^{-1} = \frac{p_0}{p_{0N}} \sqrt{\frac{\tau_{0N}}{\tau_0}}, \qquad (5.5.4)$$

and

$$G_D(v) = E \cdot v - F. \qquad (5.5.5)$$

Some of the drag functions are shown in Appendix A and Appendix B (see EBRL).
For example, for the GB528 Lapua bullet

$$G_D(v) = 0.141 \cdot v - 30.031, \qquad (5.5.6)$$

we have $E = 0.141 \cdot v$, and $F = 30.031$.

Since the distance from the muzzle to the midpoint between two screens is small, the influence of gravity in projectile trajectory is irrelevant. Thus, we can write

$$\frac{dv}{dt} = -c \cdot J^{-1} h(y) \cdot G_D(v) \cdot \frac{v}{v}. \tag{5.5.7}$$

For simplicity we consider the ICAO atmosphere. Thus $J^{-1} = 1$. The density function is also $h(y) = 1$.

From (5.5.7), we can write:

$$\frac{dv}{dx} = -\frac{c \cdot [E \cdot v - F)}{v}. \tag{5.5.8}$$

Hence

$$-\frac{v \cdot dv}{(E \cdot v - F)} = c \cdot dx. \tag{5.5.9}$$

Integrating the last equation, we have

$$E \cdot (v_0 - \bar{v}) + \ln(\frac{E \cdot v_0 - F}{E \cdot \bar{v} - F}) = c \cdot E^2 \cdot x \tag{5.5.10}$$

where v_0 is the unknown muzzle velocity, x is the distance from the bullet to the midpoint between two screens that can be measured with great accuracy, and $\bar{v}$ is the measured average velocity (5.5.2).

Solving the above equation, for example using a graphing calculator, we find the muzzle velocity.

Formula (5.5.10) can be modified to be used even when the atmosphere is not standard. In a non-standard atmosphere the equation (5.5.10) can be written:

$$E \cdot (v_0 - \bar{v}) + \ln(\frac{E \cdot v_0 - F}{E \cdot \bar{v} - F}) = c \cdot J^{-1} E^2 \cdot x \,. \qquad (5.5.11)$$

For the non-standard atmosphere the cartridges should be stored at the standard temperature (21.11 degree Celsius for the ICAO and ASM atmosphere).

The equation (5.5.11) can be used to find the muzzle velocity even when the temperature of propellant charge is the same as that of the shooting site.
The muzzle velocity is not the standard velocity of bullet.
We can study the dependence of muzzle velocity on the temperature of cartridge.

NOTE
Equation (5.5.11) can be used to measure the BC of bullet, let's say relative to the reference G_1 function. In this case, we divide the horizontal range in intervals, let's say 20 or 50 feet wide, and measure the average velocity in the middle of each interval, using chronometers.
Thus, we measure an interval BC.

Example 5.1
Find the muzzle velocity v_0 for the GB528 Lapua Scenar bullet if the average velocity, measured at the midpoint located $x = 6m$ from the muzzle of the
firearm is $\bar{v} = 826.83 m/s$.

Solution

The equation (5.5.10) for the GB528 Lapua Scenar bullet can be written:

$$0.141 \cdot (v_0 - 826.83) + \ln(\frac{0.141 \cdot v_0 - 30.031}{0.141 \cdot (826.83) - 30.031}) =$$

(5.5.12)

$$= (3.796) \cdot (0.141)^2 \cdot (6)$$

Using a graphing calculator, we find that the initial velocity is $v_0 = 829.93 m/s$.

5.6 Improved Euler's Method Applied in Exterior Ballistics
Solution of Differential Equations of Projectile Flight Using Large Steps

In Exterior Ballistics the numerical integration of differential equations of point-mass projectile trajectory, in general, is done using the numerical methods and employing a relatively small step size.

According to Robert McCoy, Euler's method used to solve the point-mass differential equations of projectile trajectory is inaccurate, while the Improved Euler's method is very accurate[32].

McCoy points out as well that at the US Army Ballistics Research Laboratory (Aberdeen Proving Ground, Maryland) the Improved Euler's method (IEM) was the preferred technique to solve the system of differential equations of point-mass projectile.
Note that McCoy, in his MCTRAJ Basic PC Program, uses a step size of 1 yard or 1 meter (McCoy's Modern Exterior Ballistics, page 184).

The C/C++ program IEM.exe (code in Appendix F), and the excel sheet named IEM (appendix F), demonstrate the accuracy of Improved Euler's method (IEM) to solve the Differential Equations of Point-Mass Projectile flying in presence or absence of drag.

The C/C++ program IEM.exe and the respective code is prepared by my son, Erio Klimi, who is a software developer.

Though the step size is unusually big (100 meters) the solution of system of differential equations (5.6.1) is sufficiently accurate. It seems that the truncation errors are insignificant.

An explanation is made employing the Taylor's formula with remainder.

The accuracy of IEM in solving the differential equations of projectile flying in vacuum is perfect even for huge step sizes, 100 - 1500 meters.

The truncation errors as well as the global truncation errors are zero, while the rounding errors are insignificant.

The C++ program IEM.exe, (code in Appendix F, EBRL) demonstrates the accuracy of the Improved Euler's Method in solving the differential equations[33] of point-mass projectile trajectory,

$$
\begin{cases}
\dfrac{dv_x}{dx} = -c \cdot h(y) \cdot \dfrac{G_D(v)}{v} \\[2mm]
\dfrac{dp}{dx} = -\dfrac{g}{v_x^2} \\[2mm]
\dfrac{dt}{dx} = \dfrac{1}{v_x} \\[2mm]
\dfrac{dy}{dx} = p
\end{cases}
\qquad , \qquad (5.6.1)
$$

even when the step size is very big, equal to the length of a soccer field.

In system (5.6.1), $p = \tan\alpha$, where α is the angle the projectile velocity forms with x-axis, (x, y) are the coordinates of projectile at a moment t, v is the projectile velocity (along the tangent to the trajectory, c is the ballistic coefficient (BC), $G_D(v)$ is the function of air resistance of the given projectile and $g = 9.80665 m/s^2$ is the gravity constant.

The component of velocity along x-axis is $v_x = v \cdot \cos\alpha$, while $v_y = v \cdot \sin\alpha$ is the vertical component. Projectile departure angle, at $x = 0$ is α_0.
A series of characteristic $G_D(v)$ functions of air resistance, for some bullets, are given in Appendix B (EBRL).

Truncation Errors for the Improved Euler's Method
We consider the trajectory of the **Lapua Scenar GB528 19.44 g (300 gr.)** bullet launched in ICAO atmosphere at an angle $\alpha_0 = 1.1471°$ with velocity $v_0 = 830 m/s$

The ballistic coefficient of the given Lapua bullet and the characteristic drag function[34] are respectively $c = 3.796 m^2/kg$ and

$$G_D(v) = 0.141 \cdot v - 30.031. \qquad (5.6.2)$$

The density function $h(y)$ at the sea level is 1, i.e. $h(y) = 1$.

The accuracy of the Improved Euler's Method (IEM) is verified comparing the data obtained using the Improved Euler's method (C++ program IEM.exe and IEM excel sheet) and the data in table 6.1, obtained using Doppler radar measurements[35].

Table 6.1. Lapua Scenar GB528 19.44 g (300 gr) bullet

Range (m)	0	300	600	900	1,200	1,500
Velocity (m/s)	830	711	604	507	422	349
Time (s)	0	0.3918	0.8507	1.3937	2.0435	2.8276
Bullet Drop (m)	0	- 0.715	- 3.203	- 8.146	- 16.571	- 30.035

Table 6.2. Comparison of True Solution with the Approximate Numerical IEM Solution

Range (m)	600		900		1200		1500	
Velocity	604	605.23	507	506.49	422	420.27	349	349.45
Time (s)	0.8507	0.847	1.394	1.3894	2.0435	2.041	2.8276	2.8272
y-coord.	8.811	8.861	9.875	9.979	7.457	7.612	0	0.1867
Bullet Drop	3.203	3.153	8.146	8.042	16.571	16.416	30.035	29.850

In table 6.2, for comparison, are given the corresponding data obtained solving system (5.6.1) by employing the improved Euler's method demonstrated in excel spreadsheet IEM.xls attached (Appendix F, EBLR).

From table 6.2, we can see that the error (difference between the true solution and the approximate one) obtained using IEM for the y-coordinate of bullet at 1500 meters is

$$e_{14} = 0.00 - 0.1867 = -0.1867m.$$

That means that the bullet, at 1500 m, will pass around 19 cm over the center of target. This approximate result is very accurate in the practice of long range shooting with small arms.

At the range 300 meters the error,

$$e_2 = 5.292 - 5.306 = -0.014m,$$

is insignificant.

Let's estimate the local truncation errors resulting when we use the Improved Euler's method (IEM).

The y-coordinate as a function of range can be expressed using Taylor formula with remainder,

$$y_{n+1} = y_n + y'(x_n) \cdot \frac{h}{1!} + y''(x) \cdot \frac{h^2}{2!} + y^{(3)}(c) \frac{h^3}{3!}, \quad x_n < c < x_{n+1} \quad (5.6.3)$$

where $h = x_{n+1} - x_n$ is the step size.

The sum of the first three terms in (5.6.3) represents the Improved Euler's formula, while the fourth term is the remainder. The remainder is equal to the local truncation error (for the interval $[x_n, x_{n+1}]$) that occurs since we approximate the true solution using limited terms of Taylor series (three terms when we use improved Euler's method).
In other words, we cut off (truncate) a part of the Taylor series solution in $[x_n, x_{n+1}]$ that is estimated using the remainder.

For the projectile trajectory that is described by the system of differential equations (5.6.1) the third derivative[36] of y with respect to x is

$$y^{(3)}(x) = -2g \cdot c \cdot (\frac{G_D(v)/v}{v^3 \cos^3(\alpha)}). \quad (5.6.4)$$

The remainder in (5.6.3), for the given Lapua bullet (using equation (5.6.2)) is

$$R_n = -2(9.80665) \cdot (3.796) \cdot [\frac{(0.141v_c - 30.031)/v_c}{v_c^3 \cos^3(\alpha_c)}] \frac{h^3}{3!}. \quad (5.6.5)$$

Hence, we can write:

$$R_n = -37.22604 \cdot [\frac{(0.141v_c - 30.031)/v_c}{v_c^3 \cos^3(\alpha_c)}] \frac{h^3}{3}. \quad (5.6.6)$$

The remainder (5.6.6), in absolute value, is not greater than the error obtained substituting in (5.6.6) instead of v_c and angle α_c respectively the value of velocity and the value of angle at the end, x_{n+1}, of the interval $[x_n, x_{n+1}]$.

For example, using the data presented in spreadsheet file EulerMethod.xls (velocity $v_c = 349.13615$, cell E143, and $v_c \cos(\alpha_c) = 349.3579$, cell E139), for the interval (1400m, 1500m), we find that the local truncation error is not greater than the absolute value of $R_M = -0.016m$
(cell I139 in the excel spreadsheet Appendix F).
Since the global truncation error for IEM theoretically is of the order h^2, i.e. $O(h^2)$, we can say that the accuracy of the numerical integration of system 5.9.1 can be increased by decreasing the step size (though in our case it is not necessary).

Thus, using C++ program for step size 0.1, 1, or 10 we find that the global truncation error at 1500 meters is

$$e_{14} = 0.00 - 0.13 = -0.13m.$$

Decreasing the step size from 100 to 0.1, we practically have an insignificant improvement of predicted outcome.

The estimation of the truncation errors is valid in general for almost all bullets that are lunched with supersonic velocities till the ranges where the velocity remains supersonic (greater than 340m/s).
For bullets, the IEM assures a great accuracy for relatively large integration steps.

IEM for Point-Mass Projectile in Vacuum

Note that for the ballistics coefficient $c = 0$ the system of differential equations (5.6.1) describes the trajectory of a projectile flying in absence of air. BC never becomes zero
As a matter of fact, the system (5.6.1) is transformed to the system of differential equation in vacuum (equation 2.7.1) when the density of air is zero.

The solution of the system of differential equations (5.9.1) with $c = 0$ is the well known parabolic trajectory of the point mass projectile flying in vacuum,

$$y = \tan(\alpha_0)x - \frac{g}{2v_0^2 \cos^2(\alpha_0)} \cdot x^2, \qquad (5.6.7)$$

where the constant of gravity is $g = 9.0665 m / s^2$.
We can get the Improved Euler's method (IEM) solution of the system (5.6.1) in vacuum, substituting in cell F12 of the Excel spreadsheet (IEM.xls) $c = 0$ instead of $c = 3.796$.

In this case, we can verify the accuracy of the Improved Euler's method (step 100 m) comparing the obtained Excel spreadsheet results with the results that we can find using the equation of parabola (5.6.7).

Thus, for the horizontal range $x = 1500 m$, substituting in the equation of parabola (5.6.7):
$g = 9.0665 m / s^2$, departure angle $\alpha_0 = 1.1471°$ and the departure velocity of bullet $v_0 = 830 m / s$, we find the true value $y_T = 14.013966 m$.

For the same horizontal range, the spreadsheet file IEM.xls (for $c = 0$) gives the value of $y_a = 14.012648 m$ (Cell E142).

The truncation error for the Improved Euler's method ($h = 100m$) is

$$e = 14.013966 - 14.012648 = 0.00132m.$$

The error above shows that the projectile will fly $0.00132m$ below the center of target located at range 1500m.
Practically the bullet will hit the center of the target.

From the practical point of view the numerical integration of system (5.6.1), in absence of air resistance, gives a perfect solution for y.

The local truncation error is zero since the third derivative of (5.6.7) is zero.
(Note as well that for $c = 0$ the third derivative of y, given by formula (5.6.4), becomes zero).

So, the remainder of the series is zero and as result, the local truncation errors are zero.
In other words, the numerical integration of the differential equations (5.6.1) of point mass projectile in absence of air resistance is free of local truncation errors.

We can say that **the local truncation error does not depend on the step size**.
It is clear that the **global truncation error is zero** as well.

Indeed, the error (difference between the true value and the approximate value) must be zero. The small error estimated above, $e = 0.00132m$ is a round off error related with the round off errors done in every step.
It is obvious that decreasing the step size, we increase the round off errors since we increase the number of iterations.

In table 6.3 is shown the round off error of the improved Euler's method for the point mass projectile flying in vacuum (step size, 100m). The estimation is done using Excel program.

Table 6.3 Round off Errors (using Excel program)

Range	300	900	1200	1500
Round off Error	$4.6 \times 10^{-5}\,m$	$2.6 \times 10^{-4}\,m$	$5.7 \times 10^{-4}\,m$	$1.3 \times 10^{-3}\,m$

We see that the round off error increases with range, i.e. with the number of iterations (3 iterations for range 300 meters, 9 for 900m and 15 for 1500m).
The round off errors, resulting by the use of IEM, for the trajectory of projectile in vacuum (range 1500 meters, step size 100 meters), are practically zero.

Interesting outcome when $c = 0$
Executing the C++ program for c= 0, step size h: 0.1, 1, 10, 100, 300 we get practically the same predicted outcomes for any range till 1500m.

Note: The results obtained using C++ programs (step 100m) are slightly different from the results obtained using Excel program.

Strange outcome for $c = 0$ **, step size** $h = 1500$

Executing the C++ program, using a huge step size h = 1500 m, we get practically the same result as above

$$(e = 14.013966 - 14.0140 = -0.000034m).$$

An explanation of the above results is related to the fact that the truncation error, and so the global truncation error, do not depend on the step size.

We can as well affirm that the IEM (Heun's method), as a **predictor-corrector algorithm,** in the case of projectile flying in absence of drag, is in accordance with Physics interpretation that considers the point-mass projectile moving independently and at the same time **along the trajectory tangent** (direction of departure velocity; prediction) and **vertically downward** (correction), in opposite direction of y-axis, under the action of gravity, i.e. according to the vector equation

$$\vec{r}(t) = \vec{v}_0 t + \frac{\vec{g}}{2} \cdot t^2 \qquad (5.6.8)$$

where $t = x/(v_0 \cos\alpha_0)$, while $\vec{r}(t)$ is the vector-position of the bullet at time t.

Truncation Errors for the Euler's Method
The excel spreadsheet file, in spreadsheet 2, named Excel1.xls (appendix F), as well as the C++ program for step size 100 m, demonstrate the solution of system (5.6.1) for the same Lapua bullet using Euler's Method.

For the Euler's method the Taylor formula with remainder is

$$y_{n+1} = y_n + y'(x_n) \cdot \frac{h}{1!} + y''(c) \cdot \frac{h^2}{2!}, \quad \text{for } x_n < c < x_{n+1} \qquad (5.6.9)$$

where $h = x_{n+1} - x_n$ is the step size.

The sum of first two terms in (5.6.9) represents the Euler's formula, while the third term is the remainder.

For the trajectory of point mass projectile that is described by the system of differential equations (5.6.1), the second derivative of y(x) is

$$y''(x) = -\frac{g}{v^2 \cos^2(\alpha)} \cdot \qquad (5.6.10)$$

For the remainder we can write:

$$R_n = -\frac{g}{v_c^2 \cos^2(\alpha_c)} \frac{h^2}{2} \cdot \qquad x_n < c < x_{n+1} \qquad (5.6.11)$$

The remainder that represents the local truncation error, changes (increases) with the increase of shooting range (because the velocity decreases with range).
We can write:

$$y_{n+1} = y_n + \tan(\alpha_n)h - \frac{g}{2v_c^2 \cos^2(\alpha_c)} \cdot h^2 \qquad (5.6.12)$$

where α_c and v_c are the projectile velocity and the projectile angle that correspond to the point on the trajectory with abscissa c, ($x_n < c < x_{n+1}$).

The sum of the first two terms in (5.6.12) represents the Euler's formula, while the third term is the local truncation error (5.6.11). The maximum value of the local truncation error is obtained close to the point on trajectory with abscissa x_{n+1}, where the component of velocity along x-axis, $v_x = v\cos\alpha$, is the smallest on the interval $[x_n, x_{n+1}]$.

For the projectile, demonstrated in Excel sheet file, Excel1.xls, the upper boundary of the local truncation error for the last interval (1400m, 1500m) is

$$R_{14} = -\frac{9.80665}{2 \cdot (342.1396)^2} \cdot (100)^2 = -0.4189m \;.$$

(For velocity see cell C139, spreadsheet Excel1.xls).
We find that the actual truncation error (see cell C142) for the range 1500 meters is

$$e(h) = y - y* = 0.00 - 4.1276 = -4.1276m .$$

That means that the bullet will hit around 4 meters over the center of the target, completely missing it.
The Euler's method solution for the system of equation (5.6.1) is not accurate when we use a large step size of 100 meters.

The upper boundary of the truncation error for the range 300m is

$$R_2 = -\frac{9.80665}{2 \cdot (712.612)^2} \cdot (100)^2 = -0.0966m .$$

(see cell C43, velocity).
Using the data of the table on the Excel.xls spreadsheet (cell C46 and C179, bullet drop), we find that the error for the range 300 meters is

$$e = y - y* = 5.292 - 5.5650 = -0.273m ,$$

since the true radar value of y-coordinate of bullet at range 300 meter is

$$y = \tan(1.1471) \cdot 300 - 0.715 = 5.292 .$$

Drop equation:

$$\bar{y} = x \cdot \tan(\alpha_0) - PQ$$

For relatively short shooting ranges and strep size 100m, the Euler's method gives acceptable solutions for the practice of long range shooting.
Of course, we can increase the accuracy by decreasing the step size, but probably we risk to increase the round off errors.

Euler's Method for Point-Mass Projectile in Absence of Drag

The system of differential equations (5.6.1), for ballistic coefficient zero, $c = 0$, describes the trajectory of point-mass projectile in absence of air resistance. The exact solution of that system (5.6.1) is given by (5.6.7).

The truncation error for the Euler's method is

$$R_n = -\frac{g}{v_0^2 \cos^2(\alpha_0)} \frac{h^2}{2} . \qquad x_n < c < x_{n+1} \qquad (5.6.13)$$

Equation (5.6.13) shows that in each interval, $[x_n, x_{n+1}]$, there is a small but constant local truncation error, i.e.

$$R_n = -\frac{g}{v_0^2 \cos^2(\alpha_0)} \frac{h^2}{2} = -\frac{9.80665}{830^2 \cdot \cos^2(1.1471)} \cdot \frac{100}{2} = -7.12 \times 10^{-4} .$$

The true error, which is practically mainly result of propagation for x =1500 meters, is

$$e = y - y* = 14.013966 - 15.08193 = -1.068m .$$

The projectile will pass $e = 1.068m$ over the center of target, completely missing it.

Conclusions

The PC program demonstrations and the analyses of the "big" step of numerical integration is an evidence that the solution of exterior ballistics problems is accurately enough even for huge integration steps.

This contradicts the rule of small steps needed to have accurate results in numerical integration of ordinary differential equations using IEM method.

Second Part

Elements of Terminal Ballistics

6

Exterior Ballistics of Fragments

Introduction

Antipersonnel ammunitions (hand-grenades, land-mines, mortar and artillery shells, etc.) are constructed to incapacitate a large number of army personnel.

In general, any antipersonnel ammunition has a metallic body that includes High Explosive (HE) explosive charge (TNT, C-4, etc.).

The detonation of the HE charge produces a large number of fast flying metallic fragments that can hit and incapacitate the military personnel located in relatively close or great distances from the center of explosion.

In general, the shape of projectile fragments is irregular, and is too far from the aerodynamic form of bullets and artillery projectiles. Only some ammunitions and anti-personnel mines distribute, around the center of explosion, regular fragments in form of metallic spheres.

Though the projectile fragments are launched with relatively large speeds, during the detonation of the explosive charge, their speed decreases very quickly with the distance due to the enormous drag forces.
The fragments lose the incapacitating effect.

The flight of fragments to the target, as well as the problems related with the construction of fragmentation ammunitions, and their efficacy can be studied using the differential equations of projectile flight in presence of drag.

The Terminal Ballistics has applications in forensics sciences as well.

6.1 The Incapacitation Criteria

The incapacitation action (injury, or lethal action) of any antipersonnel ammunition is result of the kinetic energy that the metallic fragment transmits to the human body during the collision.
The experience of wars and experiments have shown that a projectile fragment that hits a vital part of a person, would be lethal if the projectile fragment will have a kinetic energy not less than 78 Joule.

That means that a projectile fragment that is launched during the detonation of an ammunition keeps its lethal effect at a distance where the speed of the metallic fragment of mass " m " will not be reduced (as result of drag and gravity) under the speed limit value "v_l" determined by the equation

$$\frac{m \cdot v_l^2}{2} = 78 \qquad (6.1.1)$$

Hence, we find that the smallest speed that a projectile fragment must have in order to be lethal is

$$v_l = \sqrt{\frac{156}{m}} = \frac{12.5}{\sqrt{m}} \qquad (6.1.2)$$

The limit value given by (6.1.2) restricts the lethal range, of the incapacitation of the personnel located in open space, or behind light shields, and might serve as a "criterion" for the construction of antipersonnel ammunitions.

The fragments that have kinetic energy smaller than v_l can injure or damage the personnel.

Not all lethal fragments are dangerous for the personnel because part of them fly over the personnel's head while another part hit the ground before reaching to the persons around the detonation point.

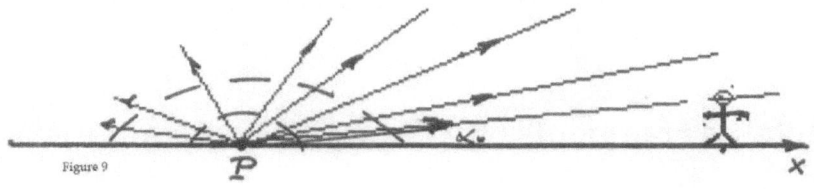

Figure 9

The flight of fragments is disordered. We cannot predict the trajectories of all fragments.

Example 1.1
Find the lethal speed limit value of a fragment of mass $3\,\text{g}$ launched during the detonation of an antipersonnel grenade.

Solution
Employing (6.1.2) we find that the speed limit is

$$v_l = \frac{12.5}{\sqrt{m}} = \frac{12.5}{\sqrt{0.003}} = 228 \text{ m/s}$$

Example 1.2
The M18 Claymore directional fragmentation mine[37] contains 700 steel balls each of mass 0.68 gram.
The casualty radius is 100 meters, while the killing radius is 50 meters.
Assuming that each fragment on average is lethal at a distance of 50 meters from the center of detonation, estimate the initial velocity of each steel ball in order that at 50 m the fragments will be lethal.

Solution
The smallest lethal speed of spherical - fragment of M18 Claymore mine is

$$v_1 = \frac{12.5}{\sqrt{m}} = \frac{12.5}{\sqrt{0.00068}} = 479 \text{ m/s}$$

6.2 Metallic Fragments of Antipersonnel Ammunitions

The projectile fragments, produced during the detonation of fragmentation ammunitions, are launched with relatively large speeds that are approximately determined by the mass of explosive charge (HE) and the mass of metallic body of ammunition.

The speed of projectile fragments decreases during the flight because of the resistance and gravity. For that reason, we need to estimate the distance (from the detonation point) where a metallic fragment of a given mass will maintain the incapacitation effect on the personnel (soft targets), or hard targets.

It is obvious that effective projectiles are those that will hit and incapacitate an average person located within a certain distance from the center of detonation of the ammunition.

A part of projectile fragments, produced during the detonation of projectile, will not hit the target.

A lethal fragment will incapacitate a person if it hits on any point of his/her body, injuring, or killing that person.
An injured person cannot continue to perform his military duties. He/she is out of action.
The lowest point of the body are the feet located on the ground level ($y=0$) while the highest point is the head, located on average $y=1.75$ meters above the ground.

The antipersonnel fragmentation ammunitions are designed to be effective in relatively short distances.
Since a person is relatively not tall (at most 1.9-2 meters), effective fragments are fragments launched with relatively small angles.

For fragments that are launched at big angles, the lethal/injury range is large. The ammunition fragments keep the injury action in relatively large ranges.

Metallic Fragments

- **Natural Irregular Fragments** (NIF) are the metallic fragments produced as results of destruction of the metallic case of the ammunition, during detonation of HE charge.
  They have an irregular form and are launched with relatively high speeds that is determined by the mass of explosive charge and the mass of the metallic body of projectile.

- **Manufactured Fragments**, usually spherical, or cuboid, are already implanted on the explosive charge.
  For example, M18 Claymore Mine, in front of HE directed to enemy, contains 700 steel balls each of mass 0.68 gram[38].

- **Secondary Fragments**, are those debris that the shock wave of explosive encounter and put in motion during the expansion.
  The range of those debris is much smaller than the range of primary fragments, i.e. natural or manufactured fragments, and so that range is not taken into consideration.

The departure velocity of natural fragments, or of manufactured fragments can be estimated using the well-known approximate method of Gurney[39].

To find the range of metallic fragments, launched during the explosion of the ammunition, we have to consider the initial velocity of fragment, the mass, and the launching angle.

The initial velocity of NIF can be approximately determined, but the mass, form, departure angle, range etc. cannot be estimated in advance.

6.3 Differential Equations of Flying Fragments

The system of differential equations that describe the flight of the fragments is[40]

$$
\begin{cases}
\dfrac{dv_x}{dx} = -c \cdot j^{-1} \cdot h(y) \cdot \dfrac{G(v)}{v} \\[2em]
\dfrac{dp}{dx} = -\dfrac{g}{v_x^2} \\[2em]
\dfrac{dt}{dx} = \dfrac{1}{v_x} \\[2em]
\dfrac{dy}{dx} = p
\end{cases}
\qquad (6.3.1)
$$

where

$g = 9.80665 \ m/s^2$ is the gravity acceleration,

v and v_x are respectively the fragment velocity and the component of velocity along x-axis,

$p = \tan(\alpha)$, α is the angle the velocity forms with the x-axis,

$$
c = i \cdot \frac{d^2}{m} \cdot 1000 \qquad\qquad (6.3.2)
$$

is the ballistic coefficient, i is the form coefficient, d is the diameter of the average cross-section area (circle), m is the mass of fragment.

The parameter

$$j^{-1} = \frac{p_0}{p_{0N}} \cdot \sqrt{\frac{\tau_{0N}}{\tau_0}} \tag{6.3.3}$$

is the scaling factor that depends on the virtual temperatures and pressures: τ_0 , τ_{0N}, p_0, p_{0N} at the firing site (τ_0, p_0), and ICAO standard conditions (τ_{0N} , p_{0N}),
The quantity

$$h(y) = (\frac{\tau_0 - 0.006328 \cdot y}{\tau_0})^{4.4} \tag{6.3.4}$$

is the density function which changes with the height y over the firing location[41], $G(v)$ is the G-function of resistance.

Approximate Solution
When the departure angle is small, we can consider the velocity equal to its x-component, i.e.

$$v \approx v_x \tag{6.3.5}$$

Thus, the first equation of the system of differential equations can be written:

$$\frac{dv}{dx} = -c \cdot j^{-1} \cdot h(y) \cdot \frac{G(v)}{v} \tag{6.3.6}$$

The $G(v)$ function of resistance can be written (see: G. Klimi, EBRM, p. 497, Xlibris 2014):

$$G(v) = 4.8110 \times 10^{-4} \cdot v^2 \cdot C_F(v/a_{0N}) \tag{6.3.7}$$

where
$\qquad C_F(v/a_{0N})$ is the drag coefficient (CD).

Substituting (6.3.7) in (6.3.6), we can write:

$$\frac{dv}{dx} = -c \cdot j^{-1} \cdot h(y) \cdot 4.8110 \times 10^{-4} \cdot v \cdot C_F \left(\frac{v}{a_{0N}}\right) = \quad (6.3.8)$$

$$= -b \cdot v$$

Where

$$b = i \cdot \frac{d^2}{m} \cdot 1000 \cdot j^{-1} \cdot h(y) \cdot 4.8110 \times 10^{-4} \cdot C_F(v/a_{0N}) \quad (6.3.9)$$

For approximate solutions[42], the Drag Coefficient is considered constant, and is called ballistic parameter, i.e.

$$C_F(v/a_{0N}) = C_F \quad\quad\quad (6.3.10)$$

The values of the form coefficient i and the ballistic parameter C_F (depend on the type of fragment: irregular, sphere, cube, etc.), are given in table 3.1.

For example, the form coefficient of a spherical fragment is $i = 1.21$, while ballistic parameter is $C_F = 0.47$ (Table 3.1)[43]. Thus, (6.3.9) can be written:

$$b = i \cdot \frac{d^2}{m} \cdot 1000 \cdot h(y) \cdot j^{-1} \cdot 4.8110 \times 10^{-4} \cdot C_F \quad\quad (6.3.11)$$

Consider approximation (6.3.5). The system of differential equations can be written:

The solution of the system of differential equations (6.3.12), for small departure angles,[44] is

$$\begin{cases} \dfrac{dv}{dx} = -b \cdot v \\[2em] \dfrac{dp}{dx} = -\dfrac{g}{v^2} \\[2em] \dfrac{dt}{dx} = \dfrac{1}{v} \\[2em] \dfrac{dy}{dx} = p \end{cases} \qquad (6.3.12)$$

Equation of Trajectory

$$y = \left(\tan(\alpha_{0T}) + \frac{g}{2bv_0^2\cos^2(\alpha_{0T})}\right) \cdot x + \frac{g(1-e^{2bx})}{4b^2v_0^2\cos^2(\alpha_{0T})} \qquad (6.3.13)$$

Velocity of Fragment $\quad v \approx v_x = v_0 \cdot \cos(\alpha_{0T}) \cdot e^{-b \cdot x}$ $\quad$ (6.3.14)

Time of Flight $\qquad t = \dfrac{e^{b \cdot x}-1}{b \cdot v_0 \cdot \cos(\alpha_{0T})}$ $\qquad$ (6.3.15)

Angle of Flight $\qquad \tan(\alpha) = \tan(\alpha_0) + \dfrac{g \cdot (1-e^{2 \cdot b \cdot x})}{2bv_0^2\cos^2(\alpha_{0T})}$ $\quad$ (6.3.16)

y-component of Velocity $\quad v_y = v_x \cdot \tan(\alpha)$ $\qquad$ (6.3.17)

where

$$b = i \cdot \frac{d^2}{m} \cdot 1000 \cdot h(y) \cdot j^{-1} \cdot 4.8110 \times 10^{-4} \cdot C_F \qquad (6.3.18)$$

Equations (6.3.13) − (6.3.16) are approximate solutions of system (6.3.1).

Table 6.1

Type fragment	Sphere Manuf	Short Cylinder	Cube Manufac.	NIF
Form Coefficient i	1.21	1.38	1.50	2.00
Parameter C_F	0.47	0.82 – 1.2	1.05	Ave 1.21

Ballistic Coefficient (BC)

Using the mass of fragment, we can find the diameter d of the average cross-section, or the diameter of the cross-section of spherical fragment.

We will consider an irregular fragment as a sphere with average diameter d of the cross-section
To find d, we can write the mass of the sphere:

$$m = \frac{4}{3} \cdot \pi \frac{d^3}{8} \rho_F = \pi \frac{d^3}{6} \rho_F = 0.526 \cdot d^3 \cdot \rho_F, \qquad (6.3.19)$$

where ρ_F is the density of the material of the fragment.

Hence, we find that the diameter of the cross-section of the spherical fragment is

$$d = 1.241 \cdot (\frac{m}{\rho_F})^{1/3}. \qquad (6.3.20)$$

We illustrate with an example the estimation of BC.

Example 3.1
Find the BC of an irregular fragment (NIF) of mass $m = 90\ g = 0.09\ kg$ and density $\rho_F = 7800\ kg/m^3$.

Solution

$$d = 1.241 \cdot \left(\frac{m}{\rho_F}\right)^{\frac{1}{3}} = 1.241 \cdot \left(\frac{0.09}{7800}\right)^{\frac{1}{3}} = 0.028 \ m.$$

In table 6.1, we have: $i = 2$.

Using (6.3.2), for the ballistic coefficient we get:

$$c = i \cdot \frac{d^2}{m} \cdot 1000 = 2 \cdot \frac{0.028^2}{0.09} \cdot 1000 = 17.47 \ m^2/kg.$$

Example 3.2
A natural irregular fragment of mass $m = 0.010 \ kg$, hit the ground at a distance $x = 100 \ m$. The initial velocity of NIF is $v_0 = 800 \ m/s$. The density of metal is $\rho_F = 7850 \ kg/m^3$.
Find the departure angle, α_0, considering the standard atmosphere on the ground, i.e. $j^{-1} = 1$, $h(y) = 1$. $i = 2$, $C_F = 1.21$.

Solution
Consider the equation of trajectory:

$$y = (\tan(\alpha_{0T}) + \frac{g}{2bcos^2(\alpha_0 T)}) \cdot x + \frac{g(1-e^{2bx})}{4b^2 v_0^2 cos^2(\alpha_0 T)}$$

To simplify the solution, we consider that the NIF is launched along x-axis at an angle $\alpha_0 = 0°$. Thus, substituting $\alpha_0 = 0$, we have:

$$\bar{y} = \frac{g}{2bv_0^2}x + \frac{g(1-e^{2bx})}{4b^2 v_0^2} \qquad (6.3.21)$$

$\bar{y}$ is the fragment drop.

Estimate b.
Average diameter of NIF is

$$d = 1.241 \cdot \left(\frac{m}{\rho_F}\right)^{\frac{1}{3}} = 1.241 \cdot \left(\frac{0.01}{7850}\right)^{\frac{1}{3}} = 0.0135 \, m.$$

BC is

$$c = i \cdot \frac{d^2}{m} \cdot 1000 = 2 \cdot \frac{0.0135^2}{0.01} \cdot 1000 = 36.2$$

$$b = 36.2 \cdot 4.8110 \times 10^{-4} \cdot 1.21 = 0.021$$

Substituting in (6.3.21), we find:

$$\bar{y} = \frac{g}{2bv_0^2}x + \frac{g(1 - e^{2bx})}{4b^2v_0^2} =$$

$$= \frac{9.80665}{2 \cdot 0.021 \cdot 800^2} \cdot 100 + \frac{9.80665 \cdot (1 - e^{2 \cdot 0.021 \cdot 100})}{4 \cdot 0.021^2 \cdot 800^2} =$$

$$= -0.534 \, m$$

Using the non-rigidity principle and equal drop model of the projectile trajectory, we can write:

$$\tan(\alpha_{0T}) = \frac{|\bar{y}|}{x} = \frac{|-0.534|}{100} = 0.00534$$

Hence, we find the departure angle:
$$\alpha_{0T} = \arctan(0.00534) = 0.306°$$

The impact velocity is

$$v \approx v_x = v_0 \cdot \cos(\alpha_{0T}) \cdot e^{-b \cdot x} = 800 \cdot \cos(0.306) \times$$

$$\times e^{-0.021 \cdot 100} = 97.96 \, m/s$$

Example 3.3
The M18 Claymore directional fragmentation mine[45]

contains 700 steel balls each of mass 0.68 gram. The casualty radius is 100 meters, while the killing radius is 50 meters.

Assuming that each fragment on average is lethal at a distance of 50 meters from the center of detonation estimate the initial velocity of each steel ball. $\rho_F = 7850 \ kg/m^3$. $i = 1.21$, $C_F = 0.47$.

Solution.

Using equation (6.1.2) we find that the smallest lethal speed of spherical- fragment of M18 is

$$v_L = \frac{12.5}{\sqrt{m}} = \frac{12.5}{\sqrt{0.00068}} = 479 \ m/s$$

We will employ the equation:

$$v \approx v_x = v_0 \cdot \cos(\alpha_{0T}) \cdot e^{-b \cdot x}$$

Let's find b.

$$d = 1.241 \cdot \left(\frac{m}{\rho_F}\right)^{\frac{1}{3}} = 1.241 \cdot \left(\frac{0.00068}{7850}\right)^{\frac{1}{3}} = 0.0055 \ m.$$

$$c = i \cdot \frac{d^2}{m} \cdot 1000 = 1.21 \cdot \frac{0.0055^2}{0.00068} \cdot 1000 = 53.65.$$

$$b = c \cdot h(y) \cdot j^{-1} \cdot 4.8110 \times 10^{-4} \cdot C_F = 53.65 \cdot 4.8110 \times 10^{-4} \cdot 0.47 = 0.0121.$$

Since departure angle is small, we have considered $\cos(\alpha_{0T}) \approx \cos(0) = 1$.

$$v \approx v_x = v_0 \cdot \cos(\alpha_{0T}) \cdot e^{-b \cdot x}$$

Substituting, $v = 479$, we get:

$$479 \approx v_0 \cdot \cos(0) \cdot e^{-0.0121 \cdot 50}$$

Hence, we find that the departure velocity is approximately:
$v_0 = 877 \ m/s$

Example 3.4.
For the Claymore mine M18 of example 3.3, find the aiming angle that assures that all fragments will hit a vertical panel of 2 meters above the ground at a distance of 50 meters considering that all 700 hundred fragments will be distributed horizontally and vertically to cover a sector 60° wide centered at the M18 mine, and two meters high.

Initial speed is $v_0 = 1000 \ m/s$.

Solution.
In example 3.4, we found:

$$b = c \cdot h(y) \cdot j^{-1} \cdot 4.8110 \times 10^{-4} C_F = 53.65 \cdot 4.8110 \times 10^{-4} \cdot 0.47$$
$$= 0.0121$$

Substituting, we find:

$$\bar{y} = \frac{g}{2bv_0^2}x + \frac{g(1-e^{2bx})}{4b^2v_0^2} == \frac{9.80665}{2 \cdot 0.0121 \cdot 1000^2} \cdot 50 + \frac{9.80665 \cdot (1-e^{2 \cdot 0.0121 \cdot 50})}{4 \cdot 0.0121^2 \cdot 1000^2}$$

$$= -0.0191m$$

$$\tan(\alpha_{0T}) = \frac{|\bar{y}|}{x} = \frac{|-0.0191|}{50} = 3.83 \times 10^{-4}$$

$$\alpha_{0T} = \arctan(3.83 \times 10^{-4}) = 0.022°$$

is the departure angle assuming the fragment hits at the ground 50 meters from the mine.

The elevation angle is:

$$\tan(E) = 2/50 = 0.04.$$

$$E = \arctan(0.04) = 2.29°.$$

The super elevation angle is

$$\alpha_{0I} = \alpha_{0T} \cdot \cos(E) = 0.022 \cdot \cos(2.29) = 0.02198°$$

Example 3.5
An Irregular fragment of mass $m = 20\ g = 0.02\ kg$, is launched during the detonation of an artillery projectile with a velocity $v_0 = 1000\ m/s$, at an angle $\alpha_0 = 1.5°$.
Density of the fragment is $\rho_F = 7800\ kg/m^3$.
Find the range of flight if the projectile detonates on the ground in standard atmosphere.

Solution
We employ the equation of the trajectory:

$$y = (\tan(\alpha_{0T}) + \frac{g}{2bv_0^2 \cos^2(\alpha_{0T})}) \cdot x + \frac{g(1-e^{2bx})}{4b^2 v_0^2 \cos^2(\alpha_{0T})}$$

We have:
$$d = 1.241 \cdot (\frac{m}{\rho_F})^{\frac{1}{3}} = 1.241 \cdot (\frac{0.020}{7800})^{\frac{1}{3}} = 0.017\ m.$$

$$c = i \cdot \frac{d^2}{m} \cdot 1000 = 2 \cdot \frac{0.017^2}{0.020} \cdot 1000 = 28.85$$

$$b = c \cdot h(y) \cdot j^{-1} \cdot 4.8110 \times 10^{-4} \cdot C_F$$
$$= 28.85 \cdot 4.8110 \times 10^{-4} \cdot 1.21 = 0.0168$$

Substituting in the equation of the trajectory and considering $y = 0$, we have:

$$0 = (\tan(1.5) + \frac{9.80665}{2 \cdot 0.0168 \cdot 1000^2 \cdot \cos^2(1.5)}) \cdot x$$

$$+ \frac{9.08665 \cdot (1 - e^{2 \cdot 0.0168 \cdot x})}{4 \cdot 0.0168^2 \cdot 1000^2 \cdot \cos^2(1.5)}$$

or

$$0.02648x + 0.008692 \cdot (1 - e^{(2 \cdot 0.0168 \cdot x)}) = 0$$

Solving for x, using a graphing calculator, we find that the range is

$$x_T = 189.25 \ m$$

Example 3.6
A spherical fragment of mass $m = 20 \ g = 0.02 \ kg$, is launched with a velocity $v_0 = 1000 \ m/s$, at an angle $\alpha_{0T} = 1.5°$.
Density of the fragment is $\rho_F = 7800 \ kg/m^3$.
Find the range of flight if the projectile detonates on the ground in standard atmosphere. ($i = 1.21$)

Solution
We employ the equation of the trajectory:

$$y = (\tan(\alpha_{0T}) + \frac{g}{2bv_0^2 \cos^2(\alpha_{0T})}) \cdot x + \frac{g(1-e^{2bx})}{4b^2 v_0^2 \cos^2(\alpha_{0T})}$$

We have:

$$d = 1.241 \cdot (\frac{m}{\rho_F})^{\frac{1}{3}} = 1.241 \cdot (\frac{0.020}{7800})^{\frac{1}{3}} = 0.017 \ m.$$

$$c = i \cdot \frac{d^2}{m} \cdot 1000 = 1.21 \cdot \frac{0.017^2}{0.020} \cdot 1000 = 17.45$$

$$b = c \cdot h(y) \cdot j^{-1} \cdot 4.8110 \times 10^{-4} \cdot C_F =$$

$$= 17.45 \cdot 4.8110 \times 10^{-4} \cdot 0.47 = 0.0039457$$

Substituting in the equation of the trajectory and considering $y = 0$, we have:

$$0 = (\tan(1.5) + \frac{9.80665}{2 \cdot 0.0039457 \cdot 1000^2 \cdot cos^2(1.5)}) \cdot x$$

$$+ \frac{9.08665 \cdot (1 - e^{2 \cdot 0.0039457 \cdot x})}{4 \cdot 0.0039457^2 \cdot 1000^2 \cdot cos^2(1.5)}$$

or $0.02743x + 0.1576 \cdot \left(1 - e^{(2 \cdot 0.0039457 \cdot x)}\right) = 0$

Solving for x, using a graphing calculator, we find that the range is $x_T = 587.67 \ m$

NOTE: Using PC program sphere.bas we obtain the same range: $x_T = 581.21 \ m$; see Exercise 4.1 section 6.4).

6.4 Solution of The System of Differential Equations

Spherical Fragment
For a spherical fragment, the G-function of resistance (see: G. Klimi, "EBLR", p. 207, Xlibris 2016) is

$$G(v) = 2.7189 \times 10^{-4} \cdot v^2 \quad for \ v \leq 1400 \ m/s \quad (6.4.1)$$

Since the $G(v)$ is a characteristic G – function of spherical fragment, the form coefficient in this case, is $i = 1$, while

$$c = i \cdot \frac{d^2}{m} \cdot 1000 = (1) \cdot \frac{d^2}{m} \cdot 1000 \qquad (6.4.2)$$

The first differential equation of (6.3.1), for the sphere, can be written:

$$\frac{dv_x}{dx} = -c \cdot j^{-1} \cdot h(y) \cdot 2.7189 \times 10^{-4} \cdot v \qquad (6.4.3)$$

All the other three differential equations of (6.3.1) remains unchanged.

Note that $G(v)$, for spherical fragments, is limited till the velocity $v \leq 1400\ m/s$.
Anyway, since for velocities of the spherical fragment that are greater than $v = 1400\ m/s$, drag function is approximately constant, we can assume that the PC program can be used even when $v > 1400\ m/s$.

Natural Irregular Fragment

According to Orlenko[46], for the natural irregular fragments, the drag coefficient (as a function of velocity v) is

$C_F(v) =$
$\begin{cases} 0.5 & for\ v \leq 150\ m/s \\ (1.49 + 0.51 \cdot \sin(860^0 - 350 \cdot \log(v)))^{-1} & 150 < v \leq 550 \\ 0.865 \cdot (1 + 50/v) & v > 550 m/s \end{cases}$
$$(6.4.4)$$

The function of resistance $G(v)$, in the first equation of system (6.3.1), is[47]:

$$G_F(v) = 4.8110 \times 10^{-4} \cdot v^2 \cdot C_F(v), \qquad (6.4.5)$$

where $C_F(v)$ is given by (6.4.4). The function G(v), presented in (6.4.5), must be substituted in the system of differential equations (6.3.1) to predict the trajectory of natural irregular fragments. Thus, for NIF the first differential equation of (6.3.1), is

$$\frac{dv_x}{dx} = -c \cdot j^{-1} \cdot h(y) \cdot 4.811 \times 10^{-4} \cdot v^2 \cdot C_F(v) \qquad (6.4.6)$$

All the other three differential equations of (6.3.1) remain unchanged.

- For the natural irregular fragments, the form coefficient can be considered $i = 2$ (see Table 6.1).

PC Programs
At the Appendix H there are shown two PC Programs in QBasic (QB64):

- Irregular.BAS,
- Sphere.BAS.

Those programs allow us to predict respectively the trajectory of irregular fragments and of spherical ones.
In each program are given examples to illustrate the use of them.

The initial velocity of the fragment launched during the detonation of explosive, could be estimated using the well-known Gurney's method.

Example 4.1 Same exercise that is solved in Exercise 3.6
A spherical fragment of mass $m = 20\,g = 0.02\,kg$, is launched with a velocity $v_0 = 1000\,m/s$, at an angle $\alpha_0 = 1.5°$.

Density of the fragment is $\rho_F = 7800\ kg/m^3$.
Find the range of flight if the projectile detonates on the ground in standard atmosphere.

Solution
PC program Sphere.Bas

Input: Initial x – coordinate = 0; Initial y – coordinate = 0; Departure angle = 1.5;
Departure Speed = 1000; Temperature of air = 15; Pressure at Firing Site = 760;
Humidity of Air = 0; Form Coefficient = 1; Mass of Spherical Fragment = 0.020
Density of Fragment = 7800; Range Wind = 0; Cross Wind = 0; Integration Step = 0.1

Results: Horizontal Range = 581.21 m; Corresponding y-coordinate = 0; Time 2.3 sec.
Terminal Speed = 100; Terminal Angle = - 5.7 degree; BC = 14.43.

Example 4.2 This exercise is Exercise 3.5.
An Irregular fragment of mass $m = 20\ g = 0.02\ kg$, is launched during the detonation of an artillery projectile with a velocity $v_0 = 1000\ m/s$, at an angle $\alpha_0 = 1.5°$.
Density of the fragment is $\rho_F = 7800\ kg/m^3$.
Find the range of flight if the projectile detonates on the ground in standard atmosphere. $i = 2$.

Solution
Using PC program Irregular.bas

Input: Initial x – coordinate = 0; Initial y – coordinate = 0; Departure angle = 1.5;

Departure Speed = 1000; Temperature of air = 15; Pressure at Firing Site = 760;
Humidity of Air = 0; Form Coefficient = 2; Mass of Spherical Fragment = 0.020
Density of Fragment = 7800; Range Wind = 0; Cross Wind = 0; Integration Step = 1

Results: Horizontal Range = 295 m; Corresponding y-coordinate = 0; Time 1.63 sec.
Terminal Speed = 74 m/s; Terminal Angle = - 5.7 degree; BC = 28.85

Example 4.3
An irregular natural steel-fragment of mass $m = 100\ g$ has an initial velocity $v_0 = 1500\ m/s$.
A WWII undetonated bomb was set to explode.
Estimate the maximum distance from the center of the controlled explosion where the fragment will hit the ground. The unexploded bomb is on the sea-level ground.

Consider: Initial velocity, $v_0 = 1500$, launching angle $\alpha_0 = 25^0$, form coefficient $i = 2$, temperature of air and temperature of propellant charge $t = 20$ degree Celsius, pressure of air at the explosion location, $p_0 = 760\ mm\ Hg$, humidity 50% = 0.5, $m = 100\ g$, no wind, density of fragment 7800 kg/m^3, integration step = 1.

Solution
Using PC program Irregular.bas.

Input: $x_0 = 0$, $y_0 = 0$, launching angle $\alpha_0 = 25^0$, initial velocity $v_0 = 1500\ m/s$, fragment mass, $m = 100\ g = 0.10\ kg$, form coefficient $i = 2$, temperature of air and temperature of propellant charge $t = 20$ degree Celsius, pressure of air at the explosion location, $p_0 = 760\ mm\ Hg$, humidity 0.5, density 7800 no wind, integration step = 1.

Results:

PC Program Irregular.bas

- For $i = 2$:
  Horizontal range, x = 826.2 m, time of flight, t = 12.65 s, terminal velocity, v = 46 m/s. terminal angle, - 74 degree, BC = 16.57.

Example 4.4
Same data as in Example 3, but the departure angle is **30, 20, 17** degree.
Use PC program Irregular.Bas.

Solution

- If the departure angle of the given fragment is **30** degree, then the horizontal range is 817 m. Terminal velocity is 47 m/s.

- If the departure angle of fragment is **20** degree then horizontal range is 822.31 m.
  Terminal velocity is 44m/s.

- If the departure angle of fragment is **17** degree then horizontal range is 812.24 m.
  Terminal velocity is 43 m/s.

Note. The maximum range is obtained when the departure angle is around 20 degree.

Example 4.5
A spherical steel-fragment of mass $m = 100\ g$ has the initial velocity $v_0 = 1500\ m/s$, launching angle $\alpha_0 = 25^0$, initial velocity

$v_0 = 1500 \, m/s$, fragment mass, $m = 100 \, g = 0.10 \, kg$, form coefficient $i = 1$, temperature of air 15 degree, pressure of air at the explosion location, $p_0 = 760 \, mm \, Hg$, humidity 50%, density of fragment, 7800, no wind, integration step = 1.

Solution
Using PC program Spheric.bas we have:

- For $i = 1$, horizontal range, x= 1667 m, time of flight, t = 18.37 s, terminal velocity, v = 62 m/s. terminal angle, - 76 degree, BC = 8.44.

6.5 Public Safe Evacuation Zone Related to Munition Disposal

We aim to predict the range of fragments of munition explosion and to give some indicators of the minimum Public Evacuation Safe Zone related to the detonation of unexploded conventional munitions.
(Edoardo Mori, Esq. Italy, is coauthor of this section)

We predict the range of natural fragments that are lunched during the detonation of unexploded munitions using the well-known model of the system of differential equations.

For public safety reasons, the reader has to consider a larger than predicted fragment range.
The readers must be cautious when they use the theoretical results presented in the book.

Drag Coefficient C_D of the Natural Irregular Fragments.

We consider unclassified paper "Fragment Hazardous Investigation Program" by Powell, J. G. ed alt, ref.[48]

The $G_F(v)$ function of resistance is related to a respective drag coefficient C_D, which is represented graphically in figure 1.

According to Powell ed alt., **"at least 95% of all artillery projectile fragments will have a C_D (Drag Coefficient) between minimum and maximum graphs shown in Fig. 1"**.

Note that minimum, average and maximum C_D curves correspond to the fragments that have respectively less, average, or maximum drag.

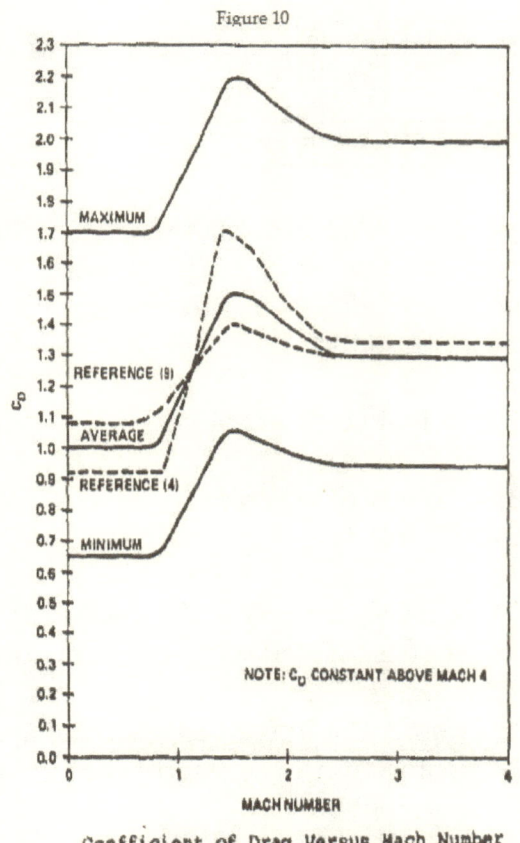

Figure 10

Coefficient of Drag Versus Mach Number

The natural fragment that corresponds to the minimum C_D curve has a greater range than all fragments that correspond to a C_D curve that is above the minimum curve.

Using the C_D curves of Fig. 2, for a typical natural fragment (of cannon 155 mm) of mass 0.0875 kg, launched with velocity 1980 m/s, at an angle 20 degree, Powell ad alt., have found the results presented in Table 6. 2.

Table 6.2

C_D curve	Range	Terminal Velocity
Minimum	850 m	43.4 8m/s
Average	610 m	36.70m/s
Maximum	396 m	27.76 m/s

Using our PC program, Irregular.Bas, for the same fragment, with mass 0.0875 kg, launched in ICAO atmosphere (*Temperature 15 degree Celsius, pressure 760 mm Hg, humidity 0%*) with velocity 1980 m/s, at an angle 20 degree, we have found

Range = 824 m, terminal velocity = 43 m/s.

The PC program Irregular.Bas, solves the system of differential equation (6.3.1)using Orlenkos CD (6.4.4).

Since our predicted range, 824 m, is less than 850 m, it means that our drag coefficient C_D curve, that is related to $G_F(v)$ function (6.4.6), is between Min. C_D and Average C_D.
Based on the above bold-text statement of Powell ed. Alt., the C_D, related to $G_F(v)$, function of resistance (6.4.6), is valid for at least 95% of all bomb and projectile fragments.
That means that the model we use to predict the trajectory of natural fragments is justified.

Predicting the Evacuation Public Safety Region
Since we don't know the Min. drag coefficient C_D (Min. curve) used by Powell ed alt., to determine the radius of that safety area, we use the PC program Irregular.bas. considering the form coefficient $i = 2$.

Example 4.3.
Using Irregular.BAS, for $i = 2$, and the data for the typical natural fragment of cannon 155 mm, i.e. of mass 0.1 kg, launched with

velocity 1000 m/s, at an angle 20 degree, as well as the temperature 15 degree Celsius, pressure 760 mm, density of fragment 7800, humidity 0.5, we find a

- Possible safe evacuation radius of $r = 815\,m < 850\,m$, terminal velocity, $v_T = 44\,m/s$, time of flight, $t = 10.88$ s.

The evacuation radius $r = 815\,m$ is 50 meters smaller than the radius 850 meters, obtained by Powell ed. Altr., using the minimum C_D curve.

We note as well that the C_D curve that corresponds to the range 815 meters is between the min C_D curve and maximum curve of figure 10.

Comment:

Seems that the maximum range $r = 850$ m, of table 6.2, calculated by Powell ed. altr., using the minimum C_D is most probably the safety range related to more than 95% of all bomb and artillery projectile fragments.

6.6 G -Function of Resistance of Irregular Fragments

The trajectory of a fragment set in motion by the detonation of an HE charge is unpredictable. There are many factors that influence the flight of a NIF, such as the mass, form, departure angle, form coefficient, as well as the Drag Coefficient (CD).

- The mass, form and departure angle of a fragment are random quantities. The Irregular fragment tumbles, there is no gyroscopic stability; the fragment flight is disordered. As such, it is not possible to predict a definite trajectory and a CD related to it.

- In near-field, where the departure angle and range of fragments are relatively small, we can estimate experimentally the flux of NIFs and the casualty probability.

Studying the safe distance of Natural Irregular Fragments of bombs and artillery projectiles, obtained and set in motion by an explosive HE charge, we have used Orlenko's Drag Coefficient model[49] .

We became aware that this CD is one of many unknown CDs between the CDmin and CDmax (Fig. 10 above).

Thus, we are uncertain which CD we need to use to determine the safety distance during the demolition of aviation bombs and artillery projectiles[50].

The contemporary exterior ballistics literature does not give a CD coefficient for "each "natural irregular fragment. Such a DC does not exist.

McCoy, in his book[51], shows the reference CD of bullets (projectiles) and spherical fragments, but not the CD of any irregular fragment.

Powell ad alt. in their paper[52] show curves of three Drag Coefficients, CDmin, CDaverage and CDmax (fig. 10), of irregular fragments of aviation bombs and artillery projectiles.

The Powell's paper does not present the CD as a numerical function of Mach number.

Using the graphic curves, it is not possible to predict the trajectory of an irregular fragment, though the authors have estimated the range and the terminal velocity that correspond to the three curves (table 6.2 above).

We assume that they use at least 3 CDs as a function of Mach number (M).

Using the data of table 6.3, and the technics shown in[53], we find the G- function of resistance and the drag coefficient that correspond to MinCurve:

1. G - Function of Resistance:

$$G(v)_{min} = -0.0082948 + 0.0003136 \cdot v + 0.0003123 \cdot v^2$$

for $v \leq 270$

$$G(v)_{min} = -28.3865 + 0.0881693 \cdot v + 0.0004007 \cdot v^2$$

for $v > 270$

Corresponding Drag Coefficient:

$$C_{Dmin} = \frac{-0.0082948 + 0.0003136 \cdot M \cdot a_{0N} + 0.0003123 \cdot M^2 \cdot a_{0N}^2}{4.811 \cdot 10^{-4} \cdot M^2 \cdot a_{0N}^2},$$

$$M \leq \frac{270}{340.30} = 0.79$$

$$C_{Dmin} = \frac{-28.3866 + 0.0881693 \cdot M \cdot a_{0N} + 0.0004007 \cdot M^2 \cdot a_{0N}^2}{4.811 \cdot 10^{-4} \cdot M^2 \cdot a_{0N}^2},$$

for $M > 0.79$

where M is the Mach number; $a_{0N} = 340.30 \ m/s$ is the velocity of sound in ICAO atmosphere.

In the same way, for **Average Curve,** we find:

1. G-Function of resistance:

$$G_A(v) = -0.0000233 + 0.0000004 \cdot v + 0.0004827 \cdot v^2$$

for $v \le 260\ m/s$

$$G_A(v) = -33.2743 + 0.123442 \cdot v + 0.0005444 \cdot v^2$$

for $v > 260\ m/s$

Corresponding Drag Coefficient

$$C_D = \frac{-0.0000233 + 0.0000004 \cdot M \cdot a_{0N} + 0.0004827 \cdot M^2 \cdot a_{0N}^2}{4.811 \cdot 10^{-4} \cdot M^2 \cdot a_{0N}^2},$$

for $M \le \dfrac{260}{340.30} = 0.764$

$$C_D = \frac{-33.2743 + 0.123442 \cdot M \cdot a_{0N} + 0.0005444 \cdot M^2 \cdot a_{0N}^2}{4.811 \cdot 10^{-4} \cdot M^2 \cdot a_{0N}^2},$$

for $M > 0.764$

For Maximum Curve we find:

2. **G-Function of Resistance**

$$G_M(v) = -0.0073136 + 0.0003369 \cdot v + 0.0008164 \cdot v^2$$

for $v \le 260\ m/s$

$$G_M(v) = -31.0805 + 0.114589 \cdot v + 0.0008864 \cdot v^2$$

for $v > 260\ \frac{m}{s}$

Corresponding Drag Coefficient

$$C_D = \frac{-0.0073136+0.0003369 \cdot M \cdot a_{0N}+0.0008164 \cdot M^2 \cdot a_{0N}^2}{4.811 \cdot 10^{-4} \cdot M^2 \cdot a_{0N}^2},$$

for $M \le \frac{260}{340.30} = 0.764$

$$C_D = \frac{-31.0805+0.114589 \cdot M \cdot a_{0N}+0.00088644 \cdot M^2 \cdot a_{0N}^2}{4.811 \cdot 10^{-4} \cdot M^2 \cdot a_{0N}^2},$$

for $M > 0.764$

Note that to each Drag Coefficient, CD, corresponds a G-function of resistance, $G(v)$.

Comment. The Powell's C_D curves of Fig. 10, are valid for Mach number from $M = 0$ to $M = 4$.

Therefore, the function of resistance $G(v)$ is valid for velocities till $v = 1,360 \ m/s$.
Nevertheless, Powell's example related with Table 6.2, considers a fragment velocity of $v = 1,980 \ m/s$.

Extracting CD from Powell's Curves

To find the drag coefficient that corresponds to each CD curve of fig. 11, we need to obtain the numerical drag coefficient as a function of Mach number, assuming that the curves are truthfully drawn.

The data in Table 6.3, extracted by Aram Bagijan (Armenia),

Table 6.3. Courtesy of A. A. Bagijan

M	CDmin	M	CDave	M	CDmax
0.01	0.65	0.02	1.00	0.02	1.7
0.19	0.65	0.14	1.00	0.2	1.7
0.37	0.65	0.33	1.00	0.38	1.7
0.55	0.65	0.51	1.00	0.56	1.7
0.74	0.65	0.69	1.00	0.75	1.7
0.88	0.68	0.85	1.03	0.87	1.75
0.97	0.75	0.93	1.09	0.95	1.81
1.06	0.80	1.01	1.16	1.04	1.87
1.16	0.87	1.10	1.22	1.13	1.93
1.25	0.93	1.16	1.26	1.22	2.00
1.34	0.99	1.23	1.32	1.29	2.06
1.44	1.05	1.30	1.38	1.38	2.12
1.62	1.05	1.39	1.44	1.46	2.18
1.78	1.02	1.49	1.50	1.63	2.19
1.94	1.00	1.66	1.49	1.77	2.15
2.12	0.98	1.81	1.46	1.92	2.11
2.29	0.96	1.96	1.42	2.08	2.07
2.47	0.95	2.11	1.38	2.23	2.04
2.64	0.95	2.26	1.34	2.4	2.01
2.82	0.95	2.36	1.31	2.58	2.00
3.01	0.95	2.54	1.30	2.75	2.00
3.19	0.94	2.72	1.30	2.94	2.00
3.38	0.95	2.90	1.30	3.11	2.00
3.54	0.94	3.08	1.30	3.29	2.00
3.72	0.94	3.25	1.30	3.47	2.00
3.89	0.94	3.44	1.30	3.65	2

From the curves of fig. 10, we found the numerical drag coefficients: CDmin, CDave. and CDmax, presented in the following table 6.3.

6.7 **Trajectory of Irregular Fragments**

Range and Terminal Velocity of NIF

To find the trajectory of NIF, we solve the system of differential, by employing the G(v) functions, (1), (3) and (5), section 6.6.

Using the Minimum Function of Resistance $G(v)_{min}$, but with form coefficient i=1.628 we get:
    Range 849.37 m, Velocity = 43 m/s.

a. Using the Average Function of Resistance and i = 1.603, we find:

    Range = 609.94 m, Terminal velocity = 35 m/s.

b. Using the max G- Function of resistance and form coefficient i = 1.603, for the same Irregular Fragment we obtain:

    Range = 396.87 m, terminal velocity = 27 m/s.

Comparing results obtained in (a), (b) and (c) with those in Table 2, we see that the CD data, we found using the graph of Powell, are correct.

Note that we found form coefficient *i* using the system of differential equations (6.3.1) and the trial and error method

Comments
According to Powell, the three CD curves in Fig. 1 show the drag coefficients of five NIF shapes. Seems that the elements of the trajectory (range, terminal velocity, time, etc.) for a given mass,

initial velocity and departure angle depend on the shape of the NIF.

Thus, for a fragment of mass m =0.0875 kg, lunched at a given angle (20 degree), with velocity 1980 m/s we expect to have a flux of fragments between the minimum range of 396 m and maximum range of 850 m.

According to our calculations (see table 7.1), the trajectory of a fragment has the max range when the launching angle is around 20 degree.

Table 6.3

Angle α_0	15°	18°	20°	23°	25°
Range x	845 m	856 m	861 m	862m	859 m

Correction
We did a small adjustment to the form coefficient of function of resistance $G(v)_{min}$, considering i = 1.602, instead of i = 1.628. The table 6.3, is corrected as shown in table 6.4.

Table 6.4

C_D curve	Range	Terminal Velocity
Minimum	861 m	43 m/s
Average	610 m	36.70m/s
Maximum	396 m	27.76 m/s

For the given fragment of mass m = 0.0875 kg, lunched at a given angle (20 degree), with velocity 1980 m/s we expect to have a flux of fragments distributed between the minimum range of 396 m and maximum range of 861 m.

PC program
We compiled a PC program in Quick Basic to solve the system of equations (6.3.1), using the function G(v) presented above.

Example 7.1.
An irregular fragment of mass $m = 0.010 \ kg$ is launched at an angle $\alpha_0 = 15°$ with velocity $v_0 = 1500 \ m/s$. The density of fragment is $\rho = 7850 \ kg/m^3$. Estimate the distribution of fragments at the horizontal plane.
At firing site: Temperature of air 15 degree Celsius, pressure 750 mm Hg, Humidity 50% = 0.5, form coefficient $i = 1.603$, no wind,

Solution
Using QBasic program, we find that the fragments will be distributed between the minimum range of 200 m and max range of 436 m.
The average impact point of fragments will be at distance of 308 meters from the launching point.

Note: If the artillery projectile will explode on the ground in vertical position, we expect a uniform distribution along each direction.

2.8 Estimation of Casualties from Fragmentation Ammunitions
Case Study: Belgium Hand Grenade PRB-423

The incapacitation of the military personnel caused by the fragmentation ammunition is attributed to the kinetic energy of metallic fragments.
It is accepted that the probable lethal effect of fragments on a random person can be achieved in few seconds if that person is hit by at least one fragment that has a kinetic energy not less then 78 Joules [54, 55].

That lethal threshold value is an important indicator for the construction of fragmentation ammunition and for the evaluation of the incapacitation action of projectile fragments which usually have random irregular forms, dimensions and masses.

Note that in practice the threshold level of the kinetic energy does not mean that the person that is hit by a fragment with kinetic energy 78 joules will pass away (be killed). The probable death is most likely to occur if the person is hit in the vital parts of the body and head.

In general, for fragmentation ammunitions, the lethal range of the blast of the explosive charge is relatively small compared to the lethal range of the fragments.
The lethal action of fragmentation ammunition is due to the high speed flying fragments.

Theoretical Considerations

To evaluate the incapacitation effect of fragmentation ammunitions on personnel we use as criteria the probability of incapacitation, i.e. the **conditional probability** $p(r)$ that a solder is incapacitated given he is being hit (at least by one fragment of not less than 78 J fragment) at a distance $r = \sqrt{x^2 + y^2}$ from the center of explosion in such a way that practically the soldier can not continue to perform his/her military activity, i.e. the person is a casualty, "dead according to the 78 J criteria, or injured".

Conditional Incapacitation Probability
The incapacitating probability is a conditional probability that shows that a random soft target will be casualty given the target is at a distance "r" from the center of explosion.
The incapacitating probability $p(r)$ is determined by a combined theoretical and experimental work for any given fragmentation

ammunition. The (conditional) incapacitating probability that a standing (prone, kneeling) soldier, in plain terrain, will be lethally hit by one fragment decreases as the distance r of the target (from the center of explosion) increases.
The decrease of incapacitating probability is due to the decrease of the density of fragments with the increase of the distance r of fragments from the center of explosion and (at the same time) due to the decrease of kinetic energy (caused by air drag).

Hence, we may assume that the lethal action of a given fragmentation projectile is practically present within a certain area D around the center of explosion of the ammunition.

Outside D we can practically consider the lethality absent
So, the (conditional) incapacitation probability function (the density function),

$$p_i = p(x, y), \qquad\qquad (6.8.1)$$

of a standing (prone, kneeling) soldier is a function of his position (x, y) relative to the center of explosion of a given fragmentation ammunition.

Incapacitation Probability

If we denote

$$w = f(x, \; y) \qquad\qquad (6.8.2)$$

the probability that a soldier will be located in the area $(dx \times dy)$ centered at (x, y) then the incapacitation probability will be

$$p_i = p(x, y) dx dy \cdot f(x, y) \qquad\qquad (6.8.3)$$

Employing the Total Probability formula, for the **incapacitation probability** P_i that a soldier (located at random within the area D) will be a casualty, we can write:

$$P_i = \iint_D p(x,y) \cdot f(x,y) dxdy. \qquad (6.8.4)$$

To compute the incapacitation probability P_i given by (6.8.4) we need to know the incapacitation function (6.8.1) and the probability function (6.8.2).

The incapacitation function (6.8.1) of a given prototype of fragmentation ammunition depends on:

- The impact velocity of the ammunition on the ground (for example artillery shell).
- The angle it strikes the ground at the moment of detonation.
- The direction of soldier with respect to x-axis,
- The position of soldier: standing, kneeling, prone, crouching, etc.

We will restrict the discussion in the cases when:

- The given fragmentation ammunition detonates at rest in a vertical position and (in average) has a regular symmetrical shape, i.e. distribution of fragments around the detonation point (origin of coordinates) does not depend on the direction of the soldier with respect to x-axis.
- The solder is in one position only: standing (prone, etc.).

In this case, the incapacitating probability function (6.8.1) depends only on the distance r of the target from the center of explosion, i.e. it is the same for all points on the circle at a given distance r.

The distribution of soldiers in terrain is difficult to be determined for each particular case in battlefield conditions; besides it varies with time and characteristics of terrain.
We will consider the case where:

- The terrain is horizontal and plain, regular and uniform, no obstacles and shelters are present.
- The infantry soldiers in standing (prone, kneeling) position are uniformly distributed around the center of explosion.

This approach allows us to consider the "location" probability f(x, y) as constant.

Thus, the probability that a soldier at the moment of explosion will be situated at random in the unit area at the point r(x, y) of the lethality region D,

$$D = \pi \cdot R^2, \tag{6.8.5}$$

Is

$$f(x, y) = \frac{1}{D} = \frac{1}{\pi \cdot R^2}, \tag{6.8.6}$$

where R is the radius of the lethality region D.

Substituting (6.8.6) in (6.8.4) we find that the incapacitation probability can be estimated using the following formula:

$$P_i = \frac{1}{D} \iint_D p(x, y) dx dy. \tag{6.8.7}$$

Incapacitation Probability in Polar Coordinates

Let us assume that the incapacitation probability function (6.8.1) does not depend on the direction of the target with respect to the x-axis but is only a function of the distance $r = \sqrt{x^2 + y^2}$ of the target location (x, y) from the origin of the coordinates, i.e. the incapacitation probability function has a radial symmetry:

$$p_i(x, y) = p(\sqrt{x^2 + y^2})$$

Considering the above incapacitation function and substituting:

$$x = r\cos\theta, \quad y = r\sin\theta$$

in (6.8.4) for the incapacitation probability we have:

$$P_i = \frac{1}{D} \iint p(r) \cdot r \cdot dr d\theta. \qquad (6.8.8)$$

Using the second formula of (6.8.8) we have:

$$P_i = \frac{1}{D} \int_0^{2\pi} d\theta \int_0^R p(r) \cdot r \cdot dr = \frac{2\pi}{D} \int_0^R r \cdot p(r) dr. \qquad (6.8.9)$$

Expected Value of Incapacitated Personnel

If inside the area D there are in average N uniformly distributed soldiers, then the mathematical expected value of casualties E_i, i.e. the average number of soldiers lethally injured, as result of the fatal action of fragmentation projectiles, is

$$E_i = N \cdot P_i = 2\pi \frac{N}{D} \int_0^R p(r) dr = 2\pi \cdot n \int_0^R r \cdot p(r) dr, \qquad (6.5.10)$$

where

$$n = N/D \qquad (6.8.11)$$

is the density of the personnel that are uniformly distributed within D.

To compute the incapacitation probability P_i given by (6.8.8) or the expected value of the casualties given by (6.8.10) we need to know the lethal probability function (6.8.1).

Hand Grenade (case study: Belgium Hand Grenade PRB-423)

As an illustration of the above results, let's estimate the probability that a soldier will be incapacitated when he is at random located around the center of explosion of the Belgium hand grenade PRB-423 (weight 0.227 kg, high explosive (HE) mass 0.063 kg, average number of fragments 870, the average mass of fragments 0.105 g.) [56].

Table 8.1 –Incapacitation probability function of PRB-423

Distance r	$r_1 = 4$	$r_2 = 6$	$r_3 = 8$	$r_4 = 10$	$r_5 = 15$	$r_6 = 20$
Incapacitat ion p(r)	0.89	0.66	0.40	0.24	0.02	0.00

In Table 8.1 is given the incapacitation probability function

$$p(r) = p(\sqrt{x^2 + y^2})$$

of Belgium grenade.

First Approach

It can be easily shown that the incapacitation probability (6.5.10) can be approximately written:

$$P_i = \frac{1}{D}\int_0^{2\pi}d\theta\int_0^R p(r)\cdot r\cdot dr = \frac{2\pi}{D}\int_0^R r\cdot p(r)dr \approx \frac{2\pi}{D}\sum \bar{r}_i p(\bar{r}_i)\Delta r_i, \quad (6.8.12)$$

where

$$\bar{r}_i = (r_{i+1}+r_i)/2, \quad \Delta r_i = (r_{i+1}-r_i)\ p(\bar{r}_i)=[p(r_{i+1})+p(r_i)]/2. \quad (6.8.13)$$

Substituting (6.8.13) in (6.8.12) we find that the incapacitation probability can be approximated by the following formula:

$$P_i \approx \frac{\pi}{D}\sum(r_{i+1}^2 - r_i^2)p(\bar{r}_i). \quad (6.8.14)$$

Using the data given in Table 8.1 we find:

$$P_i \approx \frac{\pi}{D}\sum(r_{i+1}^2-r_i^2)p(\bar{r}_i)=\frac{\pi}{D}[(4^2-0^2)\cdot(0.89+1)/2+(6^2-4^2)\cdot(0.66+0.84)/2+$$

$$+(8^2-6^2)\cdot(0.40+0.66)/2+(10^2-8^2)\cdot(0.24+0.40)/2+(15^2-10^2)\cdot(0.02+0.24)/2+$$

$$+(20^2-15^2)\cdot(0.00^2+0.02)/2=74.48\pi/D.$$

Thus, the incapacitation probability of the PRB-423 is

$$P_i \approx 74.48\pi/D = 74.48\pi/(\pi R^2) = 74.48/R^2 = 74.48/(20^2) = 0.1862 \quad (6.8.15)$$

The expected value of the casualties is

$$E_i = N\cdot P_i = N\cdot 74.48\pi/D = 74.48n\pi = 234n, \quad (6.8.16)$$

where n is the density of personnel around the center of detonation, i.e.

$$n = N/D. \qquad (6.8.17)$$

If, for example the density of personnel in the battlefield area around the point of detonation of this grenade is n = 0.01 (soldiers/m²) then the average number of casualties is

$$E_i = 234 \cdot n = 234 \cdot 0.01 = 2.34. \qquad (6.8.18$$

Note
The density of soldiers in a given area can be predicted using the norms an army uses to carry out different tasks in the battlefield.

Second Approach

Using the regression techniques, the incapacitation function of PRB-423 (table 8.1) can be presented in the following form:

$$p(r) = 0.005109r^2 - 0.179193r + 1.5265. \qquad (6.8.19)$$

Substituting in (6.8.9) and integrating (with R = 20), we find that

$$P_i = \frac{2\pi}{D} \int_0^{20} r \cdot p(r)dr = 68.424 \frac{\pi}{D}. \qquad (6.8.20)$$

Hence,

$$P_i = 68.424 \frac{\pi}{D} = \frac{68.24}{R^2} = \frac{68.24}{20^2} = 0.17. \qquad (6.8.21)$$

and the expected value of the casualties (average number of casualty soldiers) is

$$E_i = N \cdot P_i = 68.24n\pi = 68.24(0.01)\pi = 2.14 \,. \qquad (6.8.22)$$

Lethal Area, Lethal Range

For simplicity in computation of casualties in military practice it is convenient to use the model of "Lethal Area". To introduce this concept, we make use of the geometric representation of the integral on the right side of the first equation (6.8.8), i.e.

$$I = \iint\limits_{D} p(x, y)dxdy \qquad (6.8.22)$$

It is evident that at the center of explosion of the fragmentation ammunition the incapacitation probability is 1 ($p_i = 1$), (Figure 11).

Lethal area of a given fragmentation ammunition is called the region A_L, around the center of explosion, inside which the incapacitation probability is equal to 1 ($p_i = p(x, y) = 1$), whereas outside A_L the probability is considered equal to 0 (A_L is also called the area of 50% casualties).

To find the lethal area A_L we take into consideration that the right sight of (2.1) gives the total volume bounded by the probability function

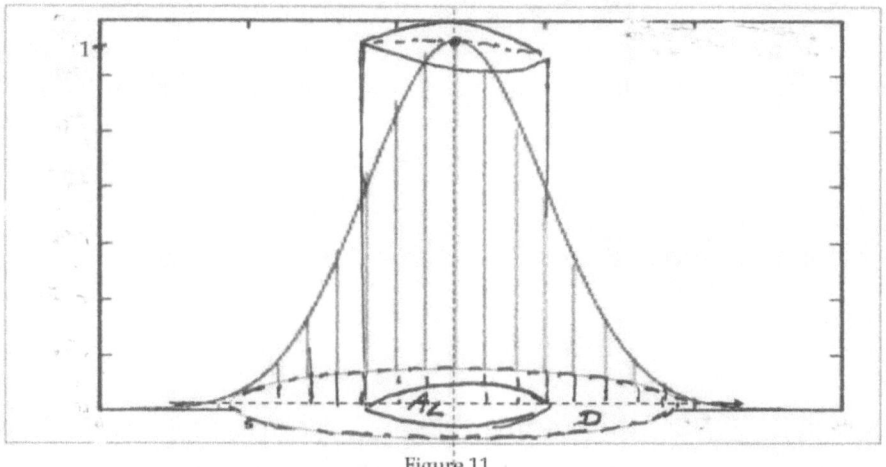

Figure 11

$$p_i = p(x, y) \qquad\qquad (6.8.22)$$

and the plane region D, where $p(0,0) = 1$.

That volume is equal to the volume inside the surface area (cylinder, parallelepiped, etc.) of the volume with base equal to the lethal area A_L and height $h = p_i = 1$. Thus, we can write:

$$(A_L \times 1) = \iint_D p(x, y)dxdy \qquad\qquad (6.8.23)$$

Using (6.8.22) and equation (6.8.23) we have:

$$A_L = P_i \cdot D. \qquad\qquad (6.8.24)$$

If the incapacitation probability functions (6.8.1) has radial symmetry ($p_i = p(r), D = \pi \cdot R_D^2$) then the lethal area also has a radial symmetry. So, we can write:

$$A_L = \pi \cdot P_i \cdot R_D^2 \qquad\qquad (6.8.25)$$

If we denote R_L the lethal range, for the lethal area we can write:

$$A_L = \pi \cdot R_L^2 \qquad\qquad (6.8.26)$$

Using (6.8.25) and (6.8.26) we find that the lethal range is

$$R_L = R_D \sqrt{P_i} \; . \qquad\qquad (6.8.27)$$

Note that R_D is the radius of the effective region of the fragmentation ammunition. In other words, it is the radius of the border where the conditional incapacitation probability becomes zero ($p_i(x_D, y_D) = 0$).

The expected number of casualties can be estimated using a simple formula:

$$E_i = N \cdot P_i = \frac{N}{D} A_L = n \cdot A_L . \qquad\qquad (6.8.28)$$

Using (6.8.26) for the case of radial symmetry, we find that the expected value of the casualties is

$$E_i = n \cdot \pi \cdot R_L^2 \qquad\qquad (6.8.29)$$

Lethal Area and Lethal Range of the Belgium Hand Grenade PRB-423

Let's apply the above results for the Belgium hand Grenade PRB-423.

Substituting P_i given by (6.8.15) in (6.5.25) and $R_D = 20m$ we find that the lethal area of the given PRB423 is

$$A_L = \pi \cdot P_i \cdot R^2 = \pi(0.1862)\cdot(20)^2 = 234m^2. \qquad (6.8.30)$$

The lethal range is

$$R_L = R\sqrt{P_i} = 20 \cdot \sqrt{0.1862} = 8.63m. \qquad (6.8.31$$

Thus, the lethal range of the hand Grenade PRB-423 is approximately 9 meters, while the lethal area is 232 square meters.

If we use the approximate probability $P_i = 0.17$ (formula 6.8.21) then the lethal range is

$$R_L = R\sqrt{P_i} = 20 \cdot \sqrt{0.17} = 8.25m. \qquad (6.8.22)$$

The obtained results are quite equal to the data given by Crevecoeur.
Thus, we can say that the method of computation used here to evaluate the *lethal area* and
lethal range is valid.

The Reduced Lethal Area

In the military handbooks the evaluation of casualties by artillery fire is based on the so called *reduced lethal area*. The reduced lethal area T of a given fragmentation ammunition is nothing else but the lethal area A_L considered as rectangular with the dimensions (a·b),

$$T = a \cdot b. \qquad (6.8.23)$$

Use of reduced lethal areas simplifies the computation of casualties from artillery fire to estimate the battery fire ammunition needed to reach a given (average) number of casualties, i.e. a given incapacitating probability, etc.

In fact, the lethal area of an artillery shell is not circular due to the fact that the artillery shell strikes the terrain with a velocity that forms an angle (terminal angle) smaller than 90 degree with the ground. The distribution of lethal fragments depends also on the departure angle of fragment during the detonation of the artillery projectile.

As an illustration, let's transform the lethal circular area A_L of the Belgium hand grenade PRB-423, calculated above:

$$A_L = \pi \cdot P_i \cdot R^2 = 234m^2 ,\qquad\qquad (6.8.24)$$

into a reduced rectangular area with base 17 meters (approximately equal to the diameter of the circular lethal area). Using (6.8.24) and (6.8.23) we can write:

$$A_L = 234 = (17) \cdot b .\qquad\qquad (6.8.25)$$

Hence, we find that $b \approx 14m$. Thus, the reduced lethal area is the rectangular area T with the dimensions a = 17, b = 14.

In table 8.2 are given some data on the *lethal area* and the *reduced lethal area* for some former Eastern European artillery ammunition caliber 85 mm and 100 mm[57]. Employing the data given in table 2 we can compute the average casualties caused by given artillery shell, a task that is very common for the artillery personnel.

As example, suppose the density of soldiers in a battlefield is n = 0.001 and the caliber of the artillery shell is 100mm:

1. Soldiers in standing position.

Using the data given in table 8.1, we find that the average number of casualties is

$$E_i = n \cdot T = 0.001 \times 410 = 0.41.$$

caliber	Angle	Soldier **standing**			Soldier in **prone**	
		Reduced area T	Reduced area (base H height)		**Reduced area T**	Reduced area (base H height)
85 mm	0 - 45°	**280 m²**	28 m	10 m	**130 m²**	19 m
100 mm	0 - 45°	**410 m²**	31 m	13.2 m	**200 m²**	22 m

2. Soldiers in prone position.
   The average number of casualties is

$$E_i = n \cdot T = 0.001 \times 200 = 0.2.$$

To cover with artillery, fire a rectangular area with dimensions 100 x 200 m², supposing a uniform distribution of soldiers in standing position we have to fire

$$N_1 = 100 \times 200 / (31 \times 13.2) = 49$$

artillery shells, caliber 100mm, in order to have an incapacitating probability 100%. That means that all 20 soldiers, $(100 \times 200 \times 0.001 = 20)$, located within the given area will be incapacitated.

A custom artillery fire task is to ensure a given lethal probability P.

Thus, for example, if the task of the artillery fire is to ensure a lethal probability P = 40% (obstructive fire) in the given area (S = 100 x 200 m²) then employing the geometric definition of probability,

$$P = S_a / S,$$

and the fact that the geometric probability does not depended on the form and location of the area S_a, inside S, that is needed to be covered by artillery fire is

$$Sa = P H S = 0.40 \times (100 \times 200) = 8000 \text{ m}^2.$$

Taking into consideration that the reduced area of the 100 mm artillery shell is T = 410 m2, it follows that we need to fire:

$$8000 / 410 = 20 \text{ artillery shells.}$$

In this case, the average number of casualties, E_i = (number of soldiers in the given area) x (lethal Probability = 40%), is

$$E_i = 20 \times 0.4 = 8.$$

The same results are obtained multiplying the density of soldiers with the 40% lethal area:

$$E_i = 0.001 \times 8000 = 8.$$

Notes on the Efficacy of Fragmentation Ammunitions

The lethal area or the lethal range of the fragmentation ammunitions are indicators of the efficacy of the fragmentation ammunitions.

Formula (2.8b) shows that the expected value of the casualties is proportional to the square of the lethal range, the effectiveness of the innovative grenades (though more expensive) is significant, for example if we compare the innovative grenades with the anti-personnel fragmentation grenades of WWII.

Thus, for example, some of the grenades or anti-personnel mines of WWII have a lethal range 4 - 5 meters. The PRB-423 is around 3 to 5 times more effective (has 3 – to 5 times more casualties) than the above mentioned grenades and has quite the same affectivity as the artillery shell 85mm (the lethal area 234 square meters of PRB-423 versus the reduced lethal area 280 square meters of 85mm artillery shell of table 2).

APPENDIX A

Reference G-functions of Resistance

1. BRL G-Functions of Resistance in ICAO Atmosphere

For the ICAO standard atmosphere, the G-function is defined by the equation:

$$G_{DA}(v) = 4.811 \times 10^{-4} \cdot v^2 C_D(v/a_{0N}), \qquad (A1)$$

where the speed of sound and the air density at the sea level are respectively:

$$a_{0N} = 340.30m/s, \quad \rho_{0N} = 1.2251kg/m^3. \qquad (A.2)$$

The density function is

$$h(y) = (\frac{288.15 - 0.006328y}{288.15})^{4.4}. \qquad (A.3)$$

Standard Function of Resistance, $G_{1A} = G_{1A}(v)$ - Type 1 Projectiles

$$G_{1A}(v) = \begin{cases} 1.0584 \times 10^{-4} \cdot v^2 & for \quad v \le 256m/s \\ 0.315754 \cdot v - 78.6769 & for \quad 256 < v \le 1000 \end{cases}, \qquad (A.4)$$

or

$$G_{1A}(v) = \begin{cases} 1.0584 \times 10^{-4} \cdot v^2 & for \quad v \le 256m/s \\ 0.331547v - 86.0227 & for \quad 256 < v \le 1200 \end{cases}. \qquad (A.5)$$

Standard Function of Resistance, $G_{2A} = G_{2A}(v)$ - Type 2 Projectiles (Usually used in Artillery Fire)

$$G_{2A}(v) = \begin{cases} 9.2868 \times 10^{-5} \cdot v^2 & for & v \le 256 m/s \\ 0.143353 \cdot v - 30.2415 & for & 256 < v \le 1300 m/s \end{cases}, \quad (A.6)$$

or

$$G_{2A}(v) = \begin{cases} 9.2868 \times 10^{-5} \cdot v^2 & for & v \le 256 m/s \\ 0.148718 \cdot v - 33.2882 & for & 256 < v \le 1700 m/s \end{cases}. \quad (A.7)$$

Standard Function of Resistance, $G_{5A} = G_{5A}(v)$ - Type 5

$$G_{5A}(v) = \begin{cases} 7.5244 \times 10^{-5} \cdot v^2 & for & v \le 256 m/s \\ 0.200207 \cdot v - 49.625 & for & 256 < v \le 1500 m/s \end{cases}, \quad (A.8)$$

or

$$G_{5A}(v) = \begin{cases} 7.5244 \times 10^{-5} \cdot v^2 & for & v \le 256 m/s \\ 0.206378 \cdot v - 53.2504 & for & 256 < v \le 1700 m/s \end{cases}. \quad (A.9)$$

Standard Function of Resistance, $G_{6A} = G_{6A}(v)$ - Type 6

$$G_{6A}(v) = \begin{cases} 1.05244 \times 10^{-4} \cdot v^2 & for & v \le 256 m/s \\ 0.142352v - 23.6937 & for & 256 < v \le 1700 m/s \end{cases}. \quad (A.10)$$

Standard Function of Resistance, $G_{7A} = G_{7A}(v)$ - Type 7

$$G_{7A}(v) = \begin{cases} 5.7679 \times 10^{-5} \cdot v^2 & for & v \le 256 m/s \\ 0.152593 \cdot v - 35.1717 & for & 256 < v \le 1700 m/s \end{cases}. \quad (A.11)$$

Standard Function of Resistance, $G_{8A} = G_{8A}(v)$ - Type 8 (Employed in artillery fire, and small arms)

$$G_{8A}(v) = \begin{cases} 1.01154 \times 10^{-4} \cdot v^2 & \textit{for} \quad v \le 256m/s \\ 0.149441 \cdot v - 29.3790 & \textit{for} \quad 256 < v \le 1500m/s \end{cases}, \quad (A.12)$$

or

$$G_{8A}(v) = \begin{cases} 1.01154 \times 10^{-4} \cdot v^2 & \textit{for} \quad v \le 256m/s \\ 0.152307 \cdot v - 31.0336 & \textit{for} \quad 256 < v \le 1700m/s \end{cases}. \quad (A.13)$$

The Spherical Projectile G-Function of Resistance,

$$G_{SA}(v) = 2.7189 \times 10^{-4} v^2, \text{ for } v \le 1400m/s. \quad (A.14)$$

According to McCoy, the roughness of surface of the sphere "causes slight drag increase at all speeds", while the very rough spheres causes a non-negligible increase of 10% of the drag at subsonic speeds and around 5% at supersonic speeds. Thus, the (A.14) can be used as well for spherical projectiles of moderate roughness.

The Cylindrical Projectile G-Function of Resistance,

$$G_{CA}(u) = 2.55078 \times 10^{-4} v^2, \text{ for } 150 < v \le 1200m/s. \quad (A.15)$$

2. BRL G-Functions of Resistance in ASM Atmosphere

In ASM standard atmosphere (the air density at the sea level, $\rho_{0N} = 1.2034kg/m^3$), the G-function is

$$G_D(v) = (4.726 \times 10^{-4})v^2 C_D(\frac{v}{a_{0N}}), \quad (A.16)$$

where $a_{0N} = 341.458m/s$ is the speed of sound at the sea level.

The density function in ASM standard atmosphere is

$$h(y) = (\frac{289.6 - 0.006328y}{289.6})^{4.4} .\qquad\qquad (A.17)$$

Standard Function of Resistance, $G_1 = G_1(v)$ - **Type 1 Projectiles**

$$G_1(v) = \begin{cases} 1.00347 \times 10^{-4} \cdot v^2 & for \quad v \le 256m/s \\ 0.312914v - 79.3976 & for \quad 256 < v \le 1000 \end{cases} (A.18)$$

or

$$G_1(v) = \begin{cases} 1.00347 \times 10^{-4} \cdot v^2 & for \quad v \le 256m/s \\ 0.332044v - 87.6845 & for \quad 256 < v \le 1200 \end{cases} .(A.19)$$

Standard Function of Resistance, $G_2 = G_2(v)$ - **Type 2 Projectiles**

$$G_2(v) = \begin{cases} 9.116233 \times 10^{-5} \cdot v^2 & for \quad v \le 256m/s \\ 0.141300 \cdot v - 29.9097 & for \quad 256 < v \le 1300m/s \end{cases} ,(A.20)$$

or

$$G_2(v) = \begin{cases} 9.116233 \times 10^{-5} \cdot v^2 & for \quad v \le 256m/s \\ 0.146467 \cdot v - 32.7991 & for \quad 256 < v \le 1700m/s \end{cases} .(A.21)$$

Standard Function of Resistance, $G_5 = G_5(v)$ - **Type 5 Projectiles**

$$G_5(v) = \begin{cases} 7.43571 \times 10^{-5} \cdot v^2 & for \quad v \le 256m/s \\ 0.19734 \cdot v - 49.0806 & for \quad 256 < v \le 1500m/s \end{cases} (A.22)$$

$$G_5(v) = \begin{cases} 7.43571 \times 10^{-5} \cdot v^2 & for \quad v \le 256m/s \\ 0.203422 \cdot v - 52.6662 & for \quad 256 < v \le 1700m/s \end{cases} . (A.23)$$

Standard Function of Resistance, $G_6 = G_6(v)$, **Type 6, Projectiles**

$$G_6(v) = \begin{cases} 1.0334 \times 10^{-4} \cdot v^2 & \textit{for} & v \le 256m/s \\ 0.140533 \cdot v - 23.6633 & \textit{for} & 256 < v \le 1700m/s \end{cases}. \qquad (A.24)$$

Standard Function of Resistance, $G_7 = G_7(v)$ - Type 7

$$G_7(v) = \begin{cases} 5.66480 \times 10^{-5} \cdot v^2 & \textit{for} & v \le 256m/s \\ 0.150355 \cdot v - 34.7319 & \textit{for} & 256m/s < v \le 1700 \end{cases}. \qquad (A.25)$$

Standard Function of Resistance, $G_8 = G_8(v)$ - Type 8

$$G_8(v) = \begin{cases} 9.9366 \times 10^{-5} \cdot v^2 & \textit{for} & v \le 256m/s \\ 0.14733 \cdot v - 29.0850 & \textit{for} & 256 < v \le 1500m/s \end{cases}, \qquad (A.26)$$

or

$$G_8(v) = \begin{cases} 9.9366 \times 10^{-5} \cdot v^2 & \textit{for} & v \le 256m/s \\ 0.150101 \cdot v - 30.6672 & \textit{for} & 256 < v \le 1700m/s \end{cases}. \qquad (A.27)$$

The Spherical Projectile G-Function of Resistance, .

$$G_S(v) = 2.60444956 \times 10^{-4} v^2. \qquad (A.28)$$

$v \le 1400m/s$

The Cylindrical Projectile G-Function of Resistance,

$$G_C(v) = 2.5117 \times 10^{-4} v^2. \qquad 150 < v \le 1200m/s \qquad (A.29)$$

3. G-Function of Resistance in TSA, ICAO, and ASM Atmosphere

In TSA standard atmosphere (the air density at the sea level, $\rho_{0N} = 1.2034 kg/m^3$), the G-function is

$$G_D(v) = 4.7320 \times 10^{-4} \cdot v^2 \cdot C_D(v/a_{0N}) , \qquad (A.30)$$

where $a_{0N} = 340.84 \ m/s$ is the speed of sound at the sea level.

The density function in ASM standard atmosphere is

$$h(y) = (\frac{289.08 - 0.006328y}{289.08})^{4.4} . \qquad (A.31)$$

The Russian G-function of Year 1943 (TSA atmosphere)

The G-function in use by the Russian army is the so-called law of resistance of 1943:

$$G_{43}(v) = \begin{cases} 7.4542 \times 10^{-5} v^2 & v \le 271 \ m/s \\ 0.187337v - 50.3858 & 271 < v \le 1400 \\ 1.2313 \times 10^{-4} v^2 & 1400 < v \le 1700 \end{cases} , (A.32)$$

or

$$G_{43c}(v) = \begin{cases} 7.4542 \times 10^{-5} v^2 & v \le 256 \\ 0.1157713v - 36.39542 & 256 < v \le 1400 \\ 1.2315 \times 10^{-4} & v > 1400 \end{cases} . (A.33)$$

The latest is used in EBNA (Xlibris, 2010).

Approximate G₄₃ - function (TSA)

$$G_{43a}(v) = \begin{cases} 7.4542 \times 10^{-5} v^2 & v \le 256 \\ 0.158738v - 36.8789 & 256 < v \le 900 \end{cases} . \qquad (A.34)$$

G₄₃ - function in ICAO atmosphere

$$G_{43I}(v) = \begin{cases} 7.535 \times 10^{-5} v^2 & v < 271 \\ 0.165114v - 39.2361 & 271 \le v \le 1020 \end{cases}. \quad \text{(A.35)}$$

G_{43} -function in ASM atmosphere is

$$G_{43A}(v) = \begin{cases} 7.4482 \times 10^{-5} v^2 & v < 271 \\ 0.163211v - 38.7839 & 271 \le v \le 1020 \end{cases}. \quad \text{(A.36)}$$

Siacci's Original Analytical Function (TSA atmosphere)

$$K_S(v) = 0.2002 \cdot v - 48.05 + [(0.1648v - 47.95)^2 + 9.6]^{1/2} + \frac{0.0442 \cdot v(v - 300)}{371 + (v/200)^{10}}.$$
(A.37)

Approximate Siacci's G-function (TSA atmosphere)

$$G_{ST}(v) = \begin{cases} 1.2094 \times 10^{-4} v^2 & v \le 257.4 \\ 0.362013v - 93.1823 & 257.4 < v \le 1200 \end{cases}. \quad \text{(A.38)}$$

Approximate Siacci's, G_{SI} - function of resistance, in ICAO atmosphere

$$G_{SI}(v) = \begin{cases} 1.2296 \times 10^{-4} v^2 & v \le 257.4 \\ 0.36874v - 95.2603 & 257.4 < v \le 1200 \end{cases}. \quad \text{(A.39)}$$

Approximate Siacci's G_{SA} - function of resistance, in ASM atmosphere

$$G_{SA}(v) = \begin{cases} 1.2078 \times 10^{-4} v^2 & v \le 257.4 \\ 0.36223v - 93.5778 & 257.4 < v \le 1200 \end{cases}. \quad \text{(A.40)}$$

Approximate Siacci's G-function, (TSA, used in EBA, Xlibris 2008)

$$K_D(v) = \begin{cases} 1.212 \cdot 10^{-4} v^2 & for & v \le 256 m/s \\ (v-240)/3 & for & 256 < v \le 900 \ m/s \end{cases} \cdot \text{(A.41)}$$

APPENDIX B
Characteristic G-functions

1. Characteristic G-function of resistance of 0.338 GB528 Lapua Scenar 19.44, 8.59 mm, velocity 830 m/s, BC = 3.796

(ICAO atmosphere):

$$G_D(v) = \begin{cases} 0.141 \cdot v - 30.031 & 325 \le v \le 850 \\ \\ 0.02438 \cdot v - 1.3696 & v < 325 \end{cases} \tag{B1}$$

(ASM atmosphere):

$$G_D(v) = \begin{cases} 0.13895 \cdot v - 29.709 & 325 \le v \le 850 \\ \\ 0.0242 \cdot v - 1.399 & v < 325 \end{cases} \tag{B2}$$

2. Characteristic G-function of resistance of 0.300 Winchester Magnum Bullet,
($m = 0.012312$ kg, caliber $d = 0.0078232$ m, departure velocity, $v_0 = 884$ m/s, form coefficient $i = 1$, ballistic coefficient $c = 4.97096$ m^2 / kg,

(ICAO atmosphere)

$$G_D(v) = \begin{cases} 6.49481 \times 10^{-5} v^2 & v < 270 \\ \\ 0.182304v - 44.271043 & v \ge 270 \end{cases} \tag{B3}$$

ASM Atmosphere

$$G_D(v) = \begin{cases} 6.3801 \times 10^{-5} v^2 & v \le 256 \\ 0.17944v - 43.592056 & v > 256 \end{cases} \tag{B4}$$

TSA Atmosphere

$$G_D(v) = \begin{cases} 6.40817 \times 10^{-5} v^2 & v \le 256 \\ 0.1798711v - 43.68047 & v > 256 \end{cases} \tag{B5}$$

3. Characteristic G-function of resistance of M118 LR bullet (Long Range, sniper bullet).: mass $m = 0.01134 kg$, caliber $d = 0.0078232\ m$, departure velocity, $v_0 = 884\ m/s$, ballistic coefficient, $BC = 5.3970\ m^2/kg$

ICAO atmosphere:

$$G_D(v) = \begin{cases} 4.81097 \times 10^{-5} v^2 & v \le 256 \\ \\ 0.178659v - 46.77305 & v > 256 \end{cases} . \tag{B6}$$

4. **G-function of Resistance of M118 Ball bullet**
The characteristic G-function of resistance of M118 Ball Bullet (Federal GM308M2. (mass $m = 0.01134\ kg$, caliber $d = 0.0078232\ m$, ballistic coefficient $c = 5.3973$, departure velocity, $v_0 = 792.48\ m/s$).

ICAO atmosphere

$$G_D(v) = \begin{cases} 6.39859 \times 10^{-5} v^2 & v \le 256 \\ \\ 0.181256v - 44.59324 & v > 256 \end{cases} . \tag{B7}$$

Note the BC of the given bullet related with G_7-function is $c = 5.9258 \ m^2 \ / \ kg$.

5. Characteristic G-function of Resistance of 300 - Grain .338 - .416 Bullet

(mass $m = 0.01944 \ kg$, caliber $d = 0.00859 \ m$, ballistic coefficient, $c = 3.7914$, departure velocity, $v_0 = 927.40 \ m/s$)

ICAO atmosphere

$$G_D(v) = \begin{cases} 7.07213 \times 10^{-5} v^2 & v \le 256 \\ \\ 0.149642v - 34.00946 & v > 256 \end{cases} \tag{B8}$$

6. G-function of Resistance of Caliber 0.30 Ball M₂ bullet

The characteristic G-function of Caliber 0.30 Ball M_2, ((mass, $m = 0.00972 \ kg$; diameter, $d = 0.0078232 \ m$; $c = 6.2965$; departure velocity, $v = 853.440 \ m/s$),

ICAO atmosphere is

$$G_D(v) = \begin{cases} 1.2027 \times 10^{-4} v^2 & v \le 256 \\ \\ 0.168240v - 35.7491 & v > 256 \end{cases} \tag{B9}$$

Average ballistic coefficient with respect to reference G_7 function:

$$c_7 = 3.7149 \text{ with respect to } G_7\text{-function } (i_7 = 0.590).$$

7. Characteristic G-function of Caliber 0.308, 168 Grain Sierra International bullet: (mass, $m = 0.010886 \ kg$; diameter,

$d = 0.0078232 \ m$; $c = 5.6325$; (i=1); muzzle velocity, $v_0 = 792.48 \ m/s$.).

ICAO atmosphere

$$G_D(v) = \begin{cases} 6.73536 \times 10^{-5} v^2 & v \le 256 \\ \\ 0.179117v - 46.77305 & v > 256 \end{cases}. \tag{B10}$$

NOTE.
The reader can find the PC programs in the book:
Klimi, G. "Elements of Exterior Ballistics: Long Range Shooting" (EBLR), Xlibris 2016.

All the PC programs can be modified to predict the trajectories of projectiles for which are known the characteristic G-functions.

Appendix G
Electronic Copies of PC Programs

Dear Reader
To request (free) electronic copies of the PC Programs that are included in the book, please send a request message to the following e-mail addresses:
iven24@aol.com.

APPENDIX H. PC Programs in QuickBasic (QB)

```
'FIND : Range,and other Elements of the Trajectory, etc.
'GIVEN: Departure Velocity, Departure Angle, Form Factor, Mass of Fragment,
Density of Fragment
'-----------------------------------------------------------
' Control DATA
' Input:
' ICAO atmosphere; x0 = y0 =0, Departure velocity =1500, departure angle 20 degree,
' Form Coefficient i = 1, Mass of Fragment m = 0.09, Density of fragment = 7850 (Iron)
' Temperature of air = 15 Celsius, Pressure = 760 mm Hg; Humidity 0 %=0,
' Range wind 0 m/s, Cross wind 0 m/s. Integration Step = 0.1

' Results: Range = 1612 m, Time of Flight = 15.69 s,
'          Terminal Speed =59 m/s, Terminal Angle = - 71 degree
'          Cross wind deflection,0 m;
'          BC = 8.7
'
'-----------------------------------------------------------------------
'Functions & Subs.

DECLARE SUB y1z1v1w1 (x, y, z, v, w, y1, z1, v1, w1, koef, pa1, wind, ys, yy, pa, ta1,
TE, De, m, Pr)
DECLARE SUB InfHyres (x0, y0, z0, v0, w0, a, h0, ta, pa, ea, tp, ta1, pa1, xx1, voo, vo1,
wind, koef, cw, vv, De, Pr, TE, m, atm, G, GA)
DECLARE SUB NPxyzvw (nk, x, x0, y, y0, z, z0, v, v0, w, w0, h, h0, k, L, r, q)
DECLARE SUB NPkoef (k, L, r, q, h, y1, z1, v1, w1)
DECLARE SUB menu (cog, cof, xf, yf, xfu, yfu, t$)
DECLARE SUB c (koef)

'Variables

DIM m(4, 4), v(4)
rendi = 4
cog = 7: cof = 0

'Zgjidhja
CLS

fillimi:
menu cog, cof, 3, 10, 21, 70, "INITIAL DATA"
```

```
InfHyres x0, y0, z0, v0, w0, a, h0, ta, pa, ea, tp, ta1, pa1, xx1, voo, vo1, wind, koef, cw,
vv, De, Pr, TE, m, atm, G, GA, Tc
c koef
F:
FOR nk = 1 TO rendi
  NPxyzvw nk, x, x0, y, y0, z, z0, v, v0, w, w0, h, h0, k, L, r, q

  y1z1v1w1 x, y, z, v, w, y1, z1, v1, w1, koef, pa1, wind, ys, yy, pa, ta1, TE, Pr
  NPkoef k, L, r, q, h, y1, z1, v1, w1
  m(nk, 1) = k: m(nk, 2) = L
  m(nk, 3) = r: m(nk, 4) = q
NEXT nk

'Calculation

FOR i = 1 TO rendi
  v(i) = 1 / 6 * (m(1, i) + 2 * m(2, i) + 2 * m(3, i) + m(4, i))
NEXT i

'New Data

x0 = x0 + h: y0 = y0 + v(1): z0 = z0 + v(2)
v0 = v0 + v(3): w0 = w0 + v(4)

IF ABS(z0) < .0001 THEN
  ymax = v0
  xmax = x0 + wind * w0
END IF

xxc = x0 + wind * w0
IF (xxc - xx1) <= .001 THEN
  xc = xxc
  yc = v0
  Tc = w0
  ac = (180 / 3.141592654#) * ATN(z0)
  vc = y0 / COS(ATN(z0))
END IF

IF x0 > 10 AND v0 <= .005 THEN

  'Display Resultst

  menu cog, cof, 6, 20, 22, 72, "RESULTS:"

  LOCATE 11, 25: PRINT "Horizontal Range [m]   = "; INT((x0 + w0 * wind - v0 / z0)
* 100 + .5) / 100
```

```
    LOCATE 12, 25: PRINT "Coresponding y-Coord [m] = "; (v0 - v0)
    LOCATE 13, 25: PRINT "Departure Angle [Deg.]    = "; INT((a) * 10000 + .5) / 10000
    LOCATE 14, 25: PRINT "Time of Flight [s]     = "; INT((w0) * 100 + .5) / 100
    LOCATE 15, 25: PRINT "Terminal Speed [m/s]    = "; INT((y0 * (1 + z0 ^ 2) ^ .5) +
.5)
    LOCATE 16, 25: PRINT "Terminal Angle [Deg.]    = "; INT((ATN(z0) * 180 /
3.141593) * 10000 + .5) / 10000
    LOCATE 17, 25: PRINT "Cross-Wind Deflection    = "; INT((cw * (w0 - x0 / (voo *
COS(a * 3.14159265# / 180)))) * 1000 + .5) / 1000
    LOCATE 18, 25: PRINT "Trajectory Vertex [m]    = "; "("; INT((xmax) * 10 + .5) / 10;
","; INT((ymax) * 100 + .5) / 100; ")"
    LOCATE 19, 25: PRINT "Ballistic Coefficient BC = "; koef
ELSE
   GOTO F:
END IF
END
SUB c (koef)
   koef = koef
END SUB

SUB InfHyres (x0, y0, z0, v0, w0, a, h0, ta, pa, ea, tp, ta1, pa1, xx1, voo, vo1, wind, koef,
cw, vv, De, Pr, TE, m, atm, G, GA, Tc)
    TE = 288.15: Pr = 760: Tc = 21
    CLS
    GOTO 400:

400
    LOCATE 5, 13: INPUT "Initial  x-coordinate of fragment    = "; x0
    LOCATE 6, 13: INPUT "Initial  y-coordinate of fragment    = "; v0
    LOCATE 7, 13: INPUT "Departure Angle [Degree]      = "; z0
    LOCATE 8, 13: INPUT "Departure Speed [m/s]       = "; y0
    LOCATE 9, 13: INPUT "Temperature of Air [C] at firing site  = "; ta
    LOCATE 10, 13: INPUT "Pressure [mm] at the firing site    = "; pa
    LOCATE 11, 13: INPUT "Humidity of Air % [decimal #]at  site  = "; ea
    LOCATE 12, 13: INPUT "Form Coefficient        = "; koef
    LOCATE 13, 13: INPUT "Mass of Fragment [kg]       = "; m
    LOCATE 14, 13: INPUT "Density of Fragment [kg/m^3       = "; De
    De = 1.241 * (m / De) ^ (1 / 3) 'Diameter of Cross Section [m]
    koef = koef * De ^ 2 * 1000 / m 'BC

    LOCATE 15, 13: INPUT "Range Wind [m/s]         = "; wind
    LOCATE 16, 13: INPUT "Cross Wind [m/s]         = "; cw
    LOCATE 17, 13: INPUT "Integration Step, 10, 1, or 0.5, 0.1 = "; h0
    vv = v0: a = z0: voo = y0
    ta = ta + 273.15
    IF ta > 273.16 AND ta <= 327.15 THEN
```

```
      ea = ea * 7.50187 * EXP(19.04 * (1 - 280.07 / ta))
   END IF
   IF ta > 255.15 AND ta < 273.15 THEN
      ea = ea * 7.50187 * EXP(22.024 * (1 - 279.24 / ta))
   END IF

   pa1 = ta / (1 - .3785 * ea / pa)
   vo1 = (voo - .4 * voo * (dm / m) + .0014 * voo * (tp - Tc))
   y0 = SQR(vo1 ^ 2 + wind ^ 2 - 2 * vo1 * wind * COS(a * 3.141592654# / 180))
   y0 = y0 * COS(a * 3.141592654# / 180)
   z0 = TAN(a * 3.141592654# / 180)
   z0 = z0 / (1 - wind / (vo1 * COS(a * 3.141592654# / 180)))
   CLS
END SUB

SUB menu (cog, cof, xf, yf, xfu, yfu, t$)

   COLOR cog, cof
   LOCATE xf - 1, yf: PRINT t$

   LOCATE xf, yf: PRINT "É" + STRING$(yfu - yf, 205) + "»";

   FOR i = xf + 1 TO xfu
      LOCATE i, yf: PRINT "º" + SPACE$(yfu - yf) + "º";
   NEXT
   LOCATE xfu + 1, yf: PRINT "È" + STRING$(yfu - yf, 205) + "¼";
END SUB

SUB NPkoef (k, L, r, q, h, y1, z1, v1, w1)

   k = h * y1: L = h * z1
   r = h * v1: q = h * w1
END SUB

SUB NPxyzvw (nk, x, x0, y, y0, z, z0, v, v0, w, w0, h, h0, k, L, r, q)

   IF nk = 1 THEN
      x = x0: y = y0: z = z0
      v = v0: w = w0: h = h0
      GOTO fund:
   END IF

   IF nk = 2 OR nk = 3 THEN
      x = x0 + (.5 * h): y = y0 + (.5 * k)
      z = z0 + (.5 * L): v = v0 + (.5 * r)
      w = w0 + (.5 * q)
```

```
      GOTO fund:
      END IF

      IF nk = 4 THEN
        x = x0 + h: y = y0 + k: z = z0 + L
        v = v0 + r: w = w0 + q
      END IF

      fund:
      END SUB

      SUB y1z1v1w1 (x, y, z, v, w, y1, z1, v1, w1, koef, pa1, wind, ys, yy, pa, ta1, TE, Pr)

      ta1 = (TE / pa1) ^ .5
      yy = y * SQR(1 + z ^ 2)
      IF yy * ta1 > 0 THEN

        y1 = -1 * koef * (pa / Pr) * ta1 * ((pa1 - .006328 * v) / pa1) ^ 4.4 * 2.7189 * 10 ^ -4 *
      (ta1 * yy) ^ 2 / yy

      END IF
      z1 = -9.80665 / y ^ 2
      v1 = z
      w1 = 1 / y
      END SUB
```

```
'                              QBasic PC Program
'                              Irregular Fragments


'FIND : Range,and other Elements of the Trajectory, etc.
'GIVEN: Departure Velocity, Departure Angle, Form Factor, Mass of Fragment,
Density of Fragment
'------------------------------------------------------------
' Control DATA
' Input:
' ICAO atmosphere; x0 = y0 =0, Departure velocity =1500, departure angle 20 degree,
' Form Coefficient i = 2, Mass of Fragment m = 0.09, Density of fragment = 7850 (Iron)
' Temperature of air = 15 Celsius, Propellant temperature 15 C, Pressure = 760 mm Hg;
Humidity 0% = 0,
' Range wind 0 m/s, Cross wind 0 m/s. Integration Step = 1

' Results: Range = 794 m, Time of Flight = 10.75 s,
'        Terminal Speed = 43 m/s, Terminal Angle = - 68 Degree
'        Cross wind deflection,0 m;
```

```
'       BC =17.40
"--------------------------------------------------------------------

'Functions & Subs.

DECLARE SUB y1z1v1w1 (x, y, z, v, w, y1, z1, v1, w1, koef, pa1, wind, ys, yy, pa, ta1,
TE, De, m, Pr)
DECLARE SUB InfHyres (x0, y0, z0, v0, w0, a, h0, ta, pa, ea, tp, ta1, pa1, xx1, voo, vo1,
wind, koef, cw, vv, De, Pr, TE, m, atm, G, GA)
DECLARE SUB NPxyzvw (nk, x, x0, y, y0, z, z0, v, v0, w, w0, h, h0, k, L, r, q)
DECLARE SUB NPkoef (k, L, r, q, h, y1, z1, v1, w1)
DECLARE SUB menu (cog, cof, xf, yf, xfu, yfu, t$)
DECLARE SUB c (koef)

'Variables
DIM m(4, 4), v(4)
rendi = 4
cog = 7: cof = 0
'Zgjidhja
CLS
fillimi:
menu cog, cof, 3, 10, 21, 70, "INITIAL DATA"
InfHyres x0, y0, z0, v0, w0, a, h0, ta, pa, ea, tp, ta1, pa1, xx1, voo, vo1, wind, koef, cw,
vv, De, Pr, TE, m, atm, G, GA, Tc
c koef

F:
FOR nk = 1 TO rendi
   NPxyzvw nk, x, x0, y, y0, z, z0, v, v0, w, w0, h, h0, k, L, r, q
   y1z1v1w1 x, y, z, v, w, y1, z1, v1, w1, koef, pa1, wind, ys, yy, pa, ta1, TE, Pr
   NPkoef k, L, r, q, h, y1, z1, v1, w1
   m(nk, 1) = k: m(nk, 2) = L
   m(nk, 3) = r: m(nk, 4) = q
NEXT nk

'Calculation
FOR i = 1 TO rendi
   v(i) = 1 / 6 * (m(1, i) + 2 * m(2, i) + 2 * m(3, i) + m(4, i))
NEXT i

'New Data
x0 = x0 + h: y0 = y0 + v(1): z0 = z0 + v(2)
v0 = v0 + v(3): w0 = w0 + v(4)

IF ABS(z0) < .0001 THEN
   ymax = v0
```

```
      xmax = x0 + wind * w0
END IF

xxc = x0 + wind * w0
IF (xxc - xx1) <= .001 THEN
   xc = xxc
   yc = v0
   Tc = w0
   ac = (180 / 3.141592654#) * ATN(z0)
   vc = y0 / COS(ATN(z0))
END IF

IF x0 > 10 AND v0 <= .005 THEN

   'Display Resultst

   menu cog, cof, 6, 20, 22, 72, "RESULTS:"

   LOCATE 11, 25: PRINT "Horizontal Range [m]    = "; INT((x0 + w0 * wind - v0 / z0)
* 100 + .5) / 100
   LOCATE 12, 25: PRINT "Coresponding y-Coord [m] = "; (v0 - v0)
   LOCATE 13, 25: PRINT "Departure Angle [Deg.]  = "; INT((a) * 10000 + .5) / 10000
   LOCATE 14, 25: PRINT "Time of Flight [s]      = "; INT((w0) * 100 + .5) / 100
   LOCATE 15, 25: PRINT "Terminal Speed [m/s]    = "; INT((y0 * (1 + z0 ^ 2) ^ .5) +
.5)
   LOCATE 16, 25: PRINT "Terminal Angle [Deg.]   = "; INT((ATN(z0) * 180 /
3.141593) * 10000 + .5) / 10000
   LOCATE 17, 25: PRINT "Cross-Wind Deflection   = "; INT((cw * (w0 - x0 / (voo *
COS(a * 3.14159265# / 180)))) * 1000 + .5) / 1000
   LOCATE 18, 25: PRINT "Trajectory Vertex [m]   = "; "("; INT((xmax) * 10 + .5) / 10;
","; INT((ymax) * 100 + .5) / 100; ")"
   LOCATE 19, 25: PRINT "Ballistic Coefficient BC = "; koef
ELSE
   GOTO F:
END IF
END

SUB c (koef)
   koef = koef
END SUB

SUB InfHyres (x0, y0, z0, v0, w0, a, h0, ta, pa, ea, tp, ta1, pa1, xx1, voo, vo1, wind, koef,
cw, vv, De, Pr, TE, m, atm, G, GA, Tc)

   TE = 288.15: Pr = 760: Tc = 21
   CLS
```

```
GOTO 400:

400
LOCATE 5, 13: INPUT "Initial  x-coordinate of fragment      = "; x0
LOCATE 6, 13: INPUT "Initial  y-coordinate of fragment      = "; v0
LOCATE 7, 13: INPUT "Departure Angle [Degree]          = "; z0
LOCATE 8, 13: INPUT "Departure Speed [m/s]             = "; y0
LOCATE 9, 13: INPUT "Temperature of Air [C] at firing site   = "; ta
LOCATE 10, 13: INPUT "Pressure [mm] at the firing site       = "; pa
LOCATE 11, 13: INPUT "Humidity of Air % [decimal #]at  site  = "; ea
LOCATE 12, 13: INPUT "Form Coefficient             = "; koef
LOCATE 13, 13: INPUT "Mass of Fragment [kg]         = "; m
LOCATE 14, 13: INPUT "Density of Fragment [kg/m^3            = "; De
De = 1.241 * (m / De) ^ (1 / 3) 'Diameter of Cross Section [m]
koef = koef * De ^ 2 * 1000 / m 'BC

LOCATE 15, 13: INPUT "Range Wind [m/s]                 = "; wind
LOCATE 16, 13: INPUT "Cross Wind [m/s]                 = "; cw


LOCATE 17, 13: INPUT "Integration Step, 10, 1, or 0.5, 0.1  = "; h0
vv = v0: a = z0: voo = y0
ta = ta + 273.15
IF ta > 273.16 AND ta <= 327.15 THEN
    ea = ea * 7.50187 * EXP(19.04 * (1 - 280.07 / ta))
END IF
IF ta > 255.15 AND ta < 273.15 THEN
    ea = ea * 7.50187 * EXP(22.024 * (1 - 279.24 / ta))
END IF

pa1 = ta / (1 - .3785 * ea / pa)
vo1 = voo
y0 = SQR(vo1 ^ 2 + wind ^ 2 - 2 * vo1 * wind * COS(a * 3.141592654# / 180))
y0 = y0 * COS(a * 3.141592654# / 180)
z0 = TAN(a * 3.141592654# / 180)
z0 = z0 / (1 - wind / (vo1 * COS(a * 3.141592654# / 180)))
CLS
END SUB

SUB menu (cog, cof, xf, yf, xfu, yfu, t$)

COLOR cog, cof
LOCATE xf - 1, yf: PRINT t$

LOCATE xf, yf: PRINT "É" + STRING$(yfu - yf, 205) + "»";
```

```
        FOR i = xf + 1 TO xfu
            LOCATE i, yf: PRINT "º" + SPACE$(yfu - yf) + "º";
        NEXT
        LOCATE xfu + 1, yf: PRINT "È" + STRING$(yfu - yf, 205) + "¼";
    END SUB

    SUB NPkoef (k, L, r, q, h, y1, z1, v1, w1)

        k = h * y1: L = h * z1
        r = h * v1: q = h * w1
    END SUB

    SUB NPxyzvw (nk, x, x0, y, y0, z, z0, v, v0, w, w0, h, h0, k, L, r, q)

        IF nk = 1 THEN
            x = x0: y = y0: z = z0
            v = v0: w = w0: h = h0
            GOTO fund:
        END IF

        IF nk = 2 OR nk = 3 THEN
            x = x0 + (.5 * h): y = y0 + (.5 * k)
            z = z0 + (.5 * L): v = v0 + (.5 * r)
            w = w0 + (.5 * q)
            GOTO fund:
        END IF

        IF nk = 4 THEN
            x = x0 + h: y = y0 + k: z = z0 + L
            v = v0 + r: w = w0 + q
        END IF

        fund:
    END SUB

    SUB y1z1v1w1 (x, y, z, v, w, y1, z1, v1, w1, koef, pa1, wind, ys, yy, pa, ta1, TE, Pr)

        ta1 = (TE / pa1) ^ .5
        yy = y * SQR(1 + z ^ 2)

        IF yy > 550 THEN

            y1 = -1 * koef * (pa / Pr) * ta1 * ((pa1 - .006328 * v) / pa1) ^ 4.4 * .865 * 4.811 * 10 ^ -
    4 * (1 + 50 / yy) * yy

        ELSEIF yy > 150 AND yy <= 550 THEN
```

y1 = -1 * koef * (pa / Pr) * ta1 * ((pa1 - .006328 * v) / pa1) ^ 4.4 * 4.811 * 10 ^ -4 * (1.49 + 0.51 * SIN(860 * 3.14159 / 180 - 350 * LOG(yy) * 3.14159 / 180)) ^ -1 * yy

ELSEIF yy < 150 THEN

y1 = -1 * koef * (pa / Pr) * ta1 * ((pa1 - .006328 * v) / pa1) ^ 4.4 * 4.811 * 10 ^ -4 * 0.5 * yy

END IF
z1 = -9.80665 / y ^ 2
v1 = z
w1 = 1 / y
END SUB

REFERENCES

1. Carlucci, D. E., Jacobson, S. S., Ballistics: Theory and Design of Guns and Ammunition, 2nd edition, CRC Press, 2014.
2. Cranz, C., Becker, K., Exterior Ballistics, London, 1921
3. Cronander, H.A.N. S., G-Dragfunctions.xls"; G-Dragmodels.xls, http://www.cronander.net/, November 10th, 2005.
4. De Mestre, N., The Mathematics of Projectiles in Sport, Cambridge University Press,
   1990.
5. Didion, I., Cours Elémentaire De Balistique, Paris, 1852
6. Engineering Design Handbook, Trajectories, Differential Effects, and Data for Projectiles, U.S.
   Army Materiel Command, 1963 (unclassified)
7. Dmitrievskij, A. A., Exterior Ballistics, Moscow 1972
8. Field Artillery, Volume 6, DND Canada, 1992 – http://www.scribd.com/doc/4934783/BALLISTICS-AND-AMMUNITION,
   (Web access December 20, 2009)
9. Gubinim, S. G., Gorovim, S. A. Ballistics, Handbook, http://www.ssga.ru/AllMetodMaterial/metod_mat_for_ioot/metodichki/ballistica/index.htm (web access 2008).
10. Hatcher J. S., "Hatcher's Notebook", Stackpole Books, 1962.
11. Hayden, R, Almgren , T., Thomas, K., McDonald W. T., Exterior Ballistics Explained, 5th Edition, Exterior Ballistics.com.
12. Herrmann, E. E., Exterior Ballistics, U.S. Naval Institute, The College Press, 1935.
13. Hurley, J. P., and Garrod, C., Principi Di Fisica, Zanichelli, 1986.
14. Klimi, G., Exterior Ballistics with Applications – Skydiving, Parachute Fall, Flying Fragments, 1st Edition, Xlibris, 2008.
15. Klimi, G., Exterior Ballistics with Applications: Skydiving, Parachute Fall, Flying
    Fragments, 3rd edition, Xlibris, 2011
16. Klimi, G., Exterior Ballistics of Small Arms, Xlibris, 2009.
17. Klimi, G., Exterior Ballistics: A New Approach, Xlibris, 2010
18. Klimi, G., Exterior Ballistics: The Remarkable Methods, Xlibris 2014
19. Kneubuehl, B. P., What is the maximum length of a spin stabilized projectile, January1987,
    published by http://www.researchgate.net/publication/253794843

20. Krasnov, N.F, Aerodynamic of Bodies of Revolution, Elsevier Publishing Inc., NY 1970.

21. Krčmář, Jan. PC Program Ballistica 2.2, ttp://www.balistika.cz/eng/exterior.html ,   (accessed on 6 November, 2009).

22. Krčmář, Jan, Exterior Ballistics 2.4, http://www.balistika.cz/eng/exterior.html

23. McCoy, R. L., Modern Exterior Ballistics, Schiffer Publishing Ltd., 1999.

24. McCoy, R. L.,Aerodynamic Characteristics of N Caliber .22 Long Rifle Match Ammunition, BRL 1990 (Unclassified)

25. McDonald, W., Inclined Fire, June, 2003, http://www.exteriorballistics.com/ebexplained/5th/50.cfm

26. McShane, E. J., Kelly, J. L., Reno, F., Exterior Ballistics, The University of Denver Press, 1953.

27. Marvin E. Backman, Terminal Ballistics, Research Department, Naval Weapons Center, February 1976 (Approved for Public release).

28. Miller, D. A New Rule for Estimating Rifling Twist, Precision Shooting, March, 43-48 (2005)

29. Mori, E., Balistica teorica e pratica; http://www.earmi.it/balistica , November   2009.

30. Mountain Range Table of 122mm cannon Mod. 1960 - Projectile OF-472, Ministry of Defense of Albania, Tirana 1972.

31. Mucinov, S.S., Shevcenko, N.A., Zadacnik po Osnovami Strelbi is Strelkovogo   Oruzie, 1964.

32. Norwood, John M., Comparison of Approximate Methods for Airborne Gunnery Ballistics Calculations, The University of Texas at Austin, TECHNICAL REPORT,   AFAL-TR-73-179, April 1973 (unclassified)

33. Okunev, B. H., Fundamentals of Ballistics, Vol.1, Book 2, Moscow, 1943.

34. Plostins, P., McCoy, R. Wagoner, B. A., Aeroballistics Performance of the 25mm M910 TPDS- T Range Limited N Training Projectile, BRL Aberdeen Proving Ground, Maryland, 1991   (Unclassified)

35. Range Tables of Cannon 122mm, Mod. 1960, Ministry of Defense of Albania, Tirana, 1967.

36. Rinker, R. A., Understanding Firearm Ballistics, Mulberry House Publishing, 6th Ed, 2005

37. Robinson, G., Robinson, I, The motion of an arbitrarily rotating spherical projectile and its application to ball games, Online at stacks.iop.org/PhysScr/88/018101

38. Roller, D. E., Blum, R., Fisica, Vol. 1, Zanichelli, 1984.

39. Shapiro, J. M., Vneshnaja Balistika, Oborongiz ', 1946

40. Von Wahlde, R., and Metz, D., Sniper Weapon Fire Control Error Budget

Analysis, U.S. Army Research Laboratory, Aberdeen Proving Ground, MD, August 1999

(unclassified, web access July 1st, 2012).

41. Weinacht, P., Cooper, G. R., Newill, J. F. Analytical Prediction of trajectories for

High-Velocity Direct-Fire Munitions, ARL, August 2005,

(unclassified, web access   July, 2013)

42. Whelan, P. M., Hodgson, M. J., Essential Principles of Physics, J. Murray, 1979.

43. Wikipedia, External Ballistics,

http://en.wikipedia.org/wiki/External_ballistics

(web access, October 24, 2009, December 2012, August 2013)

44. Wikipedia, http://en.wikipedia.org/wiki/QuickLOAD (web accessed, 08/28/2013)

45. Wikipedia, http://www.lapua.com/en/products/sport-shooting/centerfire-rifle/23

(Web access 12/01/2013)

46. Zill, D. G., Cullen, M. R., Differential Equations with Boundary-Value, 5th Ed., Books/Cole, 2001.

END NOTES

---

[1] (http://www.exteriorballistics.com/ebexplained/4th/54.cfm)

[2] https://en.wikipedia.org/wiki/External_ballistics, table at: Predictions of several drag resistance modelling and
   measuring methods

[3] (http://www.lapua.com/en/products/sport-shooting/centerfire-rifle/23)

[4] http://www.lapua.com/en/products/sport-shooting/centerfire-rifle/23).
lapuaproductcatalogue2012usa.pdf

[5] Klimi, G. Exterior Ballistics with Applications, 3rd ed, p. 44 - 48, Xlibris 2011.

[6] Klimi, G. Exterior Ballistics with Applications, 3rd ed, p. 44 - 48, Xlibris 2011.

[7] Klimi, G. Exterior Ballistics with Applications, 3rd edition, p. 46, Xlibris 2011

[8] McCoy, Robert L, Modern Exterior Ballistics, Firing Uphill and Downhill, p.47. Schiffer 1999.

McDonald, William T, Inclined Fire,

  http://www.exteriorballistics.com/ebexplained/article1.html.

Peters, V. J., at alt. Ballistic Ranging Methods and Systems for Inclined Shooting, US Patent

Date February 2, 2010.   http://www.stoel.com/webfiles/7654029.pdf, (Accessed 06/12/ 2015).

External Ballistics (Predictions of several drag resistance modelling and measuring methods),

https://en.wikipedia.org/wiki/External_ballistics, accessed 06/14/2015
[12] Klimi, G. (2010). Exterior Ballistics: A New Approach, p. 234, Xlibris.
G. Klimi, "Exterior Ballistics: The Remarkable Methods" p. 137 – 143, Xlibris 2014
Klimi, G., Exterior Ballistics: The Remarkable Methods", table 16, p. 112, Xlibris 2014
[13] G. Klimi, "Exterior Ballistics with Applications" chapter 1, third edition, Xlibris 2011
[14] Neville de Mestre, 'The Mathematics of Projectiles in Sport", Cambridge University Press, New York, 1990
[15] McCoy, Robert, "Modern Exterior Ballistics", page 101 and page 169, Schiffer Publishing, 1999.
[16] Formulas (1.3.7) and (1.3.8) are a courtesy of Prof. James Lewis (Marquette University) he developed using
Clausius-Clapeyron equation to approximate water vapor pressures for liquid-vapor equilibrium and solid-vapor
equilibrium respectively.
[17] Klimi, G., Exterior Ballistics with Applications, 3rd Ed. equation (8.4.11), Xlibris 2011.
[18] Klimi, G., Exterior Ballistics with Applications, 3rd edition, section 8.3, Xlibris, 2011
[19] Table 2 can be obtained using PC program QuickTarget Unlimited Lapua Edition.
(Refer to http://en.wikipedia.org/wiki/External_ballistics)
[20] Shapiro, J. M., Exterior Ballistics, p.205, Moscow 50'
[21] Jan Krčmář, Ballistica2.2, Powder Temperature and Barrel length, http://www.balistika.cz/eng/exterior.html (accessed on 6 November, 2009)
[22] Rinker, R.A. "Understanding Firearm Ballistics", p.162, 6th Edition, Mulberry House Publishing, 2005
[23] McCoy, Robert L. Modern Exterior Ballistics, p.165, Schiffer Publishing, 1999
[24] Klimi, G. Exterior Ballistics with Applications, Xlibris 2008; 3rd ed. 2011
Klimi, G. Exterior Ballistics: A New Approach, Xlibris 2010
Klimi, G, Exterior ballistics: The Remarkable Methods, Xlibris 2014.
[25] Quoted from "Ballistics and Field Artillery", Vol. 6: Ballistics and Ammunition, chapter 6 , section 1, (Variations and Corrections), DND Canada, 1992.
[26] Klimi, G., Exterior Ballistics; A New Approach, Section 1.5, Xlibris 2010
[27] http://en.wikipedia.org/wiki/External_ballistics, Accessed 12/20/2015
[28] www.Lapua.com, Lapua Special Purpose, English, web access 11/25/2012
[29] See Ref. 4.
[30] www.Lapua.com, Lapua Special Purpose, English, web access 11/25/2012.
McDonald, W.T, Almgren, T. C. The Ballistic Coefficient, 2008,
http://www.exteriorballistics.com/ebexplained/articles/the_ballistic_coefficient.pdf
[31] Michael and Amy Courtney, The Truth About Ballistic Coefficients,
http://arxiv.org/ftp/arxiv/papers/0705/0705.0389.pdf (accessed 12/26/2015)
[32] McCoy, Robert., Modern Exterior Ballistics, p.166, Schiffer Publishing 1999.
[33] Klimi, G., Exterior Ballistics: A New Approach, p. 234, Xlibris 2010.
[34] Klimi, G. Exterior Ballistics with Applications (3rd ed.), p. 112. Xlibris 2011
[35] External Ballistics, [Online]. http://en.wikipedia.org/wiki/External_ballistics, Predictions of several drag resistance modelling and measuring methods (accesed May 29, 2015)
[36] Klimi, G. "Exterior Ballistics: The Remarkable Methods, (page 192-200), Xlibris 2014.
[37] M18 Claymore, http://www.fas.org/man/dod-101/sys/land/m18-claymore.htm

[38] Klimi, G. Exterior Ballistics with Applications, p. 530 - 540, 3d edition, Xlibris 2010

[39] Joseph Backofen, The Gurney Velocity: A "Constant" Affected by Previously Unrecognized Factors ,
22ndInternational Ballistics Symposium, 2005.

[40] Klimi, G., Elements of Exterior Ballistics, p. 159, Xlibris 2016

[41] For more information the reader is recommended to read chapter 3 of "Elements of Exterior Ballistics",
Klimi. G Xlibris 2016

[42] Orlenko, L. P. (Editor), Explosion Physics, p. 182 – 185, Moscow 2004) and (Orlenko, L. P. Explosion and Impact Physics, p.249-250, Moscow 2006)

[43] Orlenko, L. P. Explosion and Impact Physics, p.249-250, Moscow 2006
Orlenko, L. P (Editor), Explosion Physics, p. 182 – 185, Moscow 2004

[44] (G. Klimi, EBA, P.229, Xlibris 2014)
10 M18 Claymore, http://www.fas.org/man/dod-101/sys/land/m18-claymore.htm

[46] Orlenko, L. P (Editor), Explosion Physics, p. 182 – 185, Moscow 2004.

[47] Klimi, G. Exterior Ballistics: The Remarkable Methods, p. 75, Equation (1.6.8), Xlibris 2014

[48]Powel, J. G., Smith, W. D., McCleskey, F. Fragment Hazardous Investigation Program, NSWC, July 1981,

[49]. L. P. Orlenko (Editor), Fizika Vzriva, 3rd Edition, Fizmatgiz 2004.

[50] Mori, E., Klimi, G. Exterior Ballistics of Fragments, Public Safe Evacuation Zone related to Munition   Disposal, preprint ResearchGate, 2020

[51] McCoy R. L. Modern Exterior Ballistics, p. 112, Schiffer Military History, 1999

[52] Powel, J. G., Smith, W. D., McCleskey, F. Fragment Hazardous Investigation Program, NSWC, July 1981,

[53] Klimi, G. Exterior Ballistics: The Remarkable Methods, p. 70 – 83, Xlibris 2014

[54] Apersonnel Weapons, SIPRI, London, 1978 (The wounding power of small projectiles)
Klimi, G., Exterior Ballistics with Applications, p. 522, Xlibris, 2008
Crevecoeur, P. ANouvelle conception de la grenade a main antipersonnel in Europe@, Rev. Int. de Defense, No. 3, 1973.